Application of Nanomaterials in Chemical Sensors and Biosensors

Jayeeta Chattopadhyay
Department of Chemistry
Amity School of Engineering and Technology
Amity University Jharkhand
Ranchi, Jharkhand
India

Nimmy Srivastava
Amity Institute of Biotechnology
Amity University Jharkhand
Ranchi, Jharkhand
India

CRC Press
Taylor & Francis Group
Boca Raton   London   New York

CRC Press is an imprint of the
Taylor & Francis Group, an **informa** business

A SCIENCE PUBLISHERS BOOK

Cover illustration provided by the first author, Jayeeta Chattopadhyay.

First edition published 2021
by CRC Press
6000 Broken Sound Parkway NW, Suite 300, Boca Raton, FL 33487-2742

and by CRC Press
2 Park Square, Milton Park, Abingdon, Oxon, OX14 4RN

ISBN: 978-0-367-44073-2 (hbk)
ISBN: 978-1-032-04613-6 (pbk)
ISBN: 978-1-003-00908-5 (ebk)

Typeset in Palatino
by Radiant Productions

Preface

Recent advancements in nanotechnology with their innovative synthetic approach has led the nanomaterials into the domain of sensing applications. This descriptive book utilizes a multi-disciplinary approach to provide extensive information about sensors and elucidates the impact of nanotechnology on development of chemical and biosensors for diversified applications. The main aim of this book is not only the inclusion of various research works, which have already been reported in literature, but also to make a potential conclusion about the mechanism behind this. This book provides an invaluable tool for both frontline researchers and academicians towards the future development of nanotechnology in sensing devices.

Contents

Chapter **1**

Nanomaterials

A Way to Chemical Sensor and Biosensor—Fundamentals

1. Introduction

Since the last century, the branch 'nanotechnology' is flourishing to a greater extent. These days a wide range of research is going directly and indirectly on nanotechnology and nanomaterials. Nanotechnology is a subject that covers the development, synthetic strategies, characterizations and applications of materials and devices, which can be modified by size and shape at the nanoscale level. In this case, "nano" stands as a keyword in every aspect of the products to be utilized, which is derived from the Greek word nanos or Latin word nanus meaning 'dwarf'. This subject is the combination of physics, chemistry, material science, solid-state and biosciences is taken into consideration. Nanomaterials are one of the most important parts in nanotechnology study in which fabrication, characterization and analysis are considered to determine the morphological structure of the materials on the nanoscale (Colvin et al. 2003, Marti et al. 1994, Martin et al. 1996, Nel et al. 2006).

The study on sensor technology is plagued by ambiguity with its definitions and terminology. This field of study studies an extraordinarily wider range of subjects in which every scientific and technical discipline is considered as an important role. The term 'sensor' and 'transducer' often have been used as synonyms. However, a transducer has been defined as a device that can provide a usable electrical output in response to a specific measured physical quantity, property or condition. Most of the time, the transducer has been termed as a sensor. There are many earlier developed sensors that converts physical measured quantity to mechanical energy; for example, pneumatic energy was utilized in fluid controls and mechanical energy for kinematic control. Although the introduction of solid-state electronics developed new opportunities for sensor development and control in which sensors intensively produced an electrical output for such applications as computer-based controlling systems, archiving/recording and visual display systems. For electrical interfacing, it is required to include the systems interface and signal conditioning features that produce an integral part of the sensing system. With progress in optical computing and information processing, a new class of sensors, which involve the transduction of energy into an optical form, is likely.

2. What is a Sensor?

A sensor differs from a transducer in which a sensor converts the received signal into an electrical form only, whereas a sensor collects information from the real condition. A transducer can only convert energy from one form to another (Khanna et al. 2012).

3. Key Sensing Features

Sensors can be applied in the detection of a vast variety of physical, chemical and biological quantities, including proteins, bacteria, chemicals, gases, light intensity, motion, position, sound and many others (Figure 1.1). The functionality of sensing probes can be explained by the conversion of the measurement with the application of a transducer into a signal which represents the quantity of interest to an observer or the external world (McGrath et al. 2013).

Each sensor type offers various levels of accuracy, sensitivity, specificity and ability to operate in different environmental conditions.

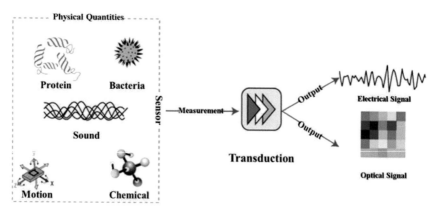

Figure 1.1: Schematic diagram of the sensing process (reproduced with permission from McGrath et al. 2013).

Before the development of sensors, researchers and industrialists also consider their cost-effectiveness. Expensive sensors are usually constituting more sophisticated features which generally display better performance characteristics. Sensing probes are used to measure the quantities in three different ways: (i) Contact, (ii) Non-contact and (iii) Sample removal.

(i) Contact

In this way, physical contact is required with the quantity of interest to be taken. Many categories are involved to sense by following the way—liquids, gaseous components and solids that contact with the human body. For obvious reasons, the application of such sensors disturbs the state of the sample or subject to some extent. The nature and the intensity of this impact are dependent on the application. Such as, comfort and biocompatibility are important aspects in the case of on-body contact sensing. In that way, sensors can cause some issues like skin irritation when it contacts for an extend period. On the other hand, fouling of the sensor can also be a major issue, therefore a process that minimizes these effects can be critical in the case of a sensor that is placed for a long duration. Moreover, these types of sensors may have restrictions on size and enclosure design. Contact sensing is commonly used in healthcare- and wellness-oriented applications, particularly where physiological measurements are required, such as in electrocardiography (ECG), electromyography (EMG) and electroencephalography (EEG). The response time of contact sensors is determined by the speed at which the quantity of interest is transported to the measurement site. For example, sensors such as ECGs that measure electrical signals have a very fast response time. In comparison, the response time of galvanic skin response (GSR) is lower as it requires the transport of sweat to an electrode, which is a slower process. Contact surface effects, such as the quality of the electrical contact between an electrode and the subject's skin, also play an important role. Poor contact can result in signal noise and the introduction of signal artifacts.

(ii) Non-contact

This type of sensing process does not need any direct contact with the quantity of interest which is to be measured. This approach has the advantage of minimum perturbation of the subject or sample. It is usually applied in natural sensing applications. These types of applications are based on the sensing probes which are ideally hidden from view and, for example, track daily activities and behaviors of individuals in their own homes. Such applications must have a minimum impact on the environment or subject of interest in order to preserve the state. Sensors that are used in non-contact modes, passive infrared (PIR), for example, generally have fast response times.

(iii) Sample Removal

This particular type of process considers an invasive collection of a representative sample through a human body or an automated sampling system. Sample removal commonly occurs in healthcare and environmental applications, to monitor *E. coli* in water or, for example, glucose levels in the blood. Such samples can be examined through either sensors or laboratory-based analytical instrumentation process. In most of the sensor-based approaches, small, hand-held disposable sensors are commonly utilized, especially when rapid measurements are needed. In this typical process, the sensor has to be placed close to the sample collection site, such as with a blood glucose sensor. Such sensors are increasingly being integrated with computing capabilities to provide sophisticated characteristics, viz., data processing, presentation, storage and remote connectivity. Contrastingly, the analytical instrumentations usually have no size limitations and typically contain a wide range of sophisticated features, such as auto-calibration or inter-sample auto-cleaning and regeneration. However, the sample has to be prepared before conducting any kind of analysis. In some instruments, sample preparation has been considered as an integrated capability. Results for nonbiological samples are generally fast and very accurate. Biological analysis, such as bacteria detection, is usually slower and can take hours or days.

4. Types of Sensors

(i) Mechanical Sensors

Mechanical sensors function on the principle of examining changes in a device or material as the result of an input that can cause mechanical deformation of that particular device or material (Fink et al. 2012). Inputs, such as motion, velocity, acceleration and displacement, results in the mechanical deformation that can be measured. When this input is converted directly into electrical output, the sensor is described as being electromechanical. Other possible output signals include magnetic, optical and thermal (Patranabis et al. 2004).

(ii) MEMS Sensors

In these types of sensors, the term MEMS is often used to describe both a type of sensing probe and their manufacturing process which can fabricate the sensor. MEMS can be defined as 3-dimensional, miniaturized mechanical and electrical structures, typically ranging from 1 to 100 mm, which can be developed using standard semiconductor manufacturing technologies. MEMS consists of mechanical microstructures, microsensors, microactuators and microelectronics, all of which are integrated onto the same silicon chip. This type of sensor is widely used in the car industry and from the early 1990s, accelerometers have been used in air-bag restraint systems, electronic stability programs (ESPs) and antilock braking systems (ABS). In recent times, the availability of inexpensive, ultra-compact, low-power multi-axis MEMS sensors has resulted in the rapid growth of customer-electronics (CE) devices. Similarly, MEMS can also be found in smartphones, tablets, game console controllers, portable gaming devices, digital cameras and camcorders. They have also found application in the healthcare domain in devices, such as blood pressure monitors, pacemakers, ventilators and respirators. Many other forms of MEMS sensors are existing in the industries and research field, two of the most important and widely used forms are accelerometers and gyroscopes, which are produced by companies such as Analog Devices and Freescale Semiconductor.

(iii) Optical Sensors

This type of sensor functions by detecting waves or photons of light, including light in the visible, infrared and ultraviolet (UV) spectral regions. They can be operated with the measurement of a change in light intensity related to light emission or absorption by a quantity of interest. They can also analyse phase changes via the application of light beams due to interaction or interference effects. In another way, the measurement of the absence or interruption of a light source can be possible. This type of sensor is commonly used in automated doors and gates to ensure no obstacles are present in their opening path. They are also used extensively in industrial applications for measuring liquids and material levels in tanks or factory production lines in the detection of the presence or absence of the objects. Optical sensors are also used with stepper-motors in applications that need position sensing and encoding, for example in automated lighting systems in the entertainment industry (Cadena et al. 2013).

(iv) Semiconductor Sensors

Semiconductor sensors have gained huge attention due to their low cost, reliability, lower power consumption, longer operational life-span and smaller form factor. This type of sensor can be obtained in a wider range of applications, including gas monitoring and pollution monitoring such as carbon monoxide, nitrogen dioxide, sulfur dioxide and ozone (Nihal et al. 2008, Wetchakun et al. 2011), breath analyzers for breath-alcohol content (BAC) measurements (Knott 2010) and domestic gas monitoring such as propane (Heberto et al. 2013). This type of sensor is popular in the detection of hydrogen, oxygen (O_2), alcohol and harmful gases, such as CO. Especially, CO detectors used for domestic purposes are one of the most popular applications in gas-monitoring semiconductors. This type of sensing probe consist of a sensing layer and sensor base and is placed inside a porous structure. In this typical structure, the sensing layer is composed of a porous, thick-film metal oxide semiconductor (MOS) layer, such as tin oxide (SnO_2) or tungsten trioxide (WO_3). This is deposited onto a micro-sensor layer containing electrodes that can measure the resistance of the sensing layer and a heater that can heat the sensing layer to 200°C to 400°C temperature. The principle of the process is that when the metal oxide is heated to a high temperature in air, oxygen is absorbed on the crystal surface during the process with a negative charge and thus donor e^- in the crystal surface is transferred to the absorbed oxygen, leaving positive charges in a space-charge layer. In this process, a potential barrier is formed against electron flow. The presence of reducing gases, such as CO or H_2, catalytic reduction at the preabsorbed oxygen layer decreases the resistance of the sensor. On the other hand, the presence of oxidizing gases, such as nitrogen dioxide and ozone gases, has the opposite effect, leading to an increase in resistance value. The magnitude of resistance change can be correlated to the concentration of the gas species.

Semiconductor temperature sensors are based on the change of voltage across a p-n junction, which exhibits strong thermal dependence. The temperature sensor is formed with a silicon diode in which the forward bias across the diode has a temperature coefficient of approximately 2.0–2.3 mV/°C. The measurements are constituted of holding the bias current constant and measuring voltage changes. For accurate readings, the sensor needs to be calibrated via a two-point calibration process. For more accurate measurements, diode-connected bipolar transistors are used. Again, a constant current is applied through the base-emitter junction, generating a voltage that is a linear function of the temperature. An offset maybe applied to convert the signal from absolute temperature to Celsius or Fahrenheit. Typically, operating ranges are –55°C to +150°C. Semiconductor temperature sensors are often categorized by their output signal type, which can be analog (voltage and current), logic or digital (Gyorki 2009). The key advantages of this sensor type are ease of integration into a circuit, general ruggedness and low cost. Their primary disadvantages are limitations of accuracy and stability, often poor thermal chip design, and slow response time (CAPGO 2010, Fraden 2010).

There are a variety of optical semiconductor sensors, the most common of which is the photodiode, a type of photodetector that converts light into either current or voltage. Photodiodes normally have a window or optical fiber connection to allow light to reach a p-n or a PIN junction (an intrinsic semiconductor region between p-type and n-type semiconductor regions). Photodiodes often use a PIN junction rather than a p-n junction to increase the speed of response. When a photon of sufficient energy strikes the depletion region of the diode, it may hit an atom with sufficient energy to release an electron, thereby creating a free electron (and a positively charged electron-hole). Free electrons and holes in the depletion region, or one diffusion length away, are pulled away in an applied electrical field. The holes move toward the anode and the electrons move toward the cathode, resulting in a photocurrent. This photocurrent is the sum of both the dark current (without light) and the light current, so the dark current must be minimized to enhance the sensitivity of the device. Photodiodes are used in a variety of applications, including pulse oximeters, blood particle analysers, nuclear radiation detectors and smoke detectors.

(v) Electrochemical Sensors

An electrochemical sensor is composed of a sensing or working electrode, a reference electrode and in many cases a counter electrode. These electrodes are typically placed in contact with either a liquid or a solid electrolyte. In the low-temperature range (< 140° C), electrochemical sensors are used to monitor pH, conductivity, dissolved ions and dissolved gases. For measurements at high temperatures (> 500° C), such as the measurement of exhaust gases and molten metals, solid electrolyte sensors are used (Guth et al. 2009). Electrochemical sensors work on the principle of measuring an electrical parameter of the sample of interest. They can be categorized based on the measurement-approach employed. Electrochemical sensors present a number of advantages, including low power consumption, high sensitivity, good accuracy and resistance to surface-poisoning effects. However, their sensitivity, selectivity and stability are highly influenced by environmental conditions, particularly temperature. Environmental conditions also have a strong influence on operational lifespan; for example, a sensor's useful life will be significantly reduced in hot and dry environments. Cross-sensitivity to other gases can be a problem for gas sensors. Oversaturation of the sensor to the species of interest can also reduce the sensor's lifespan. The potentiometric sensor measures differences in potential (voltage) between the working electrode and a reference electrode. The working electrode's potential depends on the concentration (more exactly, the ion activity) of the species of interest (Banica 2012). For example, in a pH sensor, the electric potential, created between the working electrode and the reference electrode, is a function of the pH value of the solution being measured. Other applications of potentiometric sensors include ion-selective electrodes for both inorganic (for example, monitoring of metal ion contamination in environmental samples or profiling of blood electrolytes) and organic ions (such as aromatic aldehyde or ibuprofen in human serum samples).

(vi) Biochemical Sensors

Biosensors use biochemical mechanisms to identify an analyte of interest in chemical, environmental (air, soil and water) and biological samples (blood, saliva and urine). The sensor uses an immobilized biological material, which could be an enzyme, antibody, nucleic acid or hormone in a self-contained device (Figure 1.2). The biological material being used in the biosensor device is immobilized in a manner that maintains its bioactivity. Methods utilized include membrane (for example, electroactive polymers) entrapment, physical bonding and noncovalentor covalent binding. The immobilization process results in contact being made between the immobilized biological material and the transducer. When

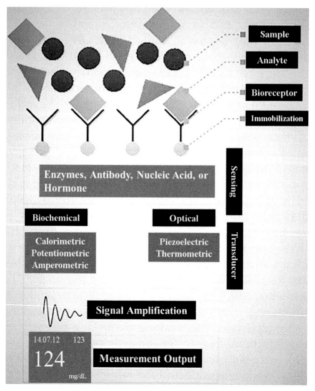

Figure 1.2: Schematic diagram of Biosensing process (Reproduced with permission from McGrath et al. 2013).

an analyte comes into contact with the immobilized biological material, the transducer produces a measurable output, such as a current, change in mass or a change in color. Indirect methods can also be utilized in which a biochemical reaction occurs between the analyte and sensor material, resulting in a product. During the reaction, measurable quantities such as heat, gas (for example, oxygen), electrons or hydrogen ions are produced and can be measured.

5. Nanomaterials and their Synthetic Strategies

Nanomaterial in the form of its studies in respect to the studies related to fabrication, characterization and analysis of materials with morphological features on the nanoscale in at least one dimension has been taken the main part of nanotechnology (Colvin et al. 2003, Martin et al. 1994, Martin et al. 1996, Nel et al. 2006, Oberdarster et al. 2005). The nanoscale is usually considered as the particle size which is smaller than 100 nm. However, many a time, it is extended to a dimension lesser than 1 mm. In recent times, the European Commission considered the definition of a nanomaterial as a natural, incidental or manufactured material containing particles in an unbound state, as an aggregate or as an agglomerate and where, for 50% or more of the particles in the number size distribution, one or more external dimensions is in the size range 1–100 nm.

Nanomaterials can be classified on the basis of dimension (D) of their features into 0D, 1D, 2D and 3D nano-structures (Pokropivny et al. 2008). The 0-dimensional nanomaterials are further be categorized into the following groups, viz., nanoparticles, nanospheres, quantum dots, isolated molecules and atoms, which can be represented as point structures with nanoscale dimensions (Chan et al. 2002, Pashchanka et al. 2010, Sun et al. 2006). Similarly, one-dimensional structures can be classified as nanotubes and nanowires with non-nanoscale only in one dimension (Pashchanka et al. 2010, Hillhouse et al. 2011, Zhao et al. 2009, Chopra et al. 2007). On the other hand, when the dimension is two, the nanomaterials can be presented in the following categories, viz., nanosheet, nanoplates, nanobelts and nanodisc, which present in nanoscale structure with nanoscale in one dimension (Wang et al. 2011, Yi et al. 2009, Chen et al. 2011, Ciesielski et al. 2010). Most importantly, 3D nanostructures, viz., nanotetrapods, nanoflowers and nanocombs have arbitrary structures; of course with nanoscale features in any of three dimensions (Lu et al. 2008, Song et al. 2011). The nanostructured materials can be formed with a large classification of functional materials, which includes metals, metal oxides, ionic compounds, ceramics, semiconductors, insulators, organics, polymers, biological materials, bioorganisms and many more. Carbon is an element that can be an efficient example of all dimensionalities, viz., 0D fullerene (hollow buckyball) (Ivanovskii et al. 2003), 1D carbon nanotubes (CNTs) (Wang et al. 2005, Chen et al. 2003, Lam et al. 2004, Jeong et al. 2009), 2D graphene (Wang et al. 2011) and 3D graphite nanostructures. Not only carbon but also a wide range of nanomaterials with different dimensions of metal (Sondi et al. 2004, Colaianni et al. 2009), metal oxide (Pashchanka et al. 2010, Chen et al. 2011, Patzke et al. 2002, Chen et al. 2007, Sayle et al. 2009), semiconductor (Pietryga et al. 2008, Chen et al. 2010, Fang et al. 2006, Madsen et al. 2011, Okamoto et al. 2011), organic (Zhao et al. 2009, Forey et al. 2005, Zhao et al. 2008), polymers (Barbero et al. 2010), biomaterials (Katz et al. 2004, Liu et al. 2011, George et al. 2011, Lo et al. 2010) and their composites (George et al. 2011, Arami et al. 2011, Bao et al. 2010, Chen et al. 2009) are available in the literature. Different forms of nano-materials can be developed or fabricated using many different methodologies. Usually, nanomaterials are synthesized by three main processes, viz., top-down, bottom-up and combination (Zhang et al. 2003, Gasparotto et al. 2012). In the first approach, bulk starting materials will be broken down into nanoscale structures by various methods, such as photolithographic patterning, wet etching, plasma etching, reactive-ion etching, laser processing, electrochemical etching and grinding (Chen et al. 2010, Bang et al. 2007). The approach can be used for the production of nanoparticles, nanorod and nanowires of metal oxide, semiconductor, metal and polymer materials. The main advantages of these methods include well-controlled parameters and large-scale manufacturability. However, they suffer from high material loss, relatively high cost and slow production rate.

The most potential aspect of all the nanomaterials lies in their special characteristics associated with nanoscale dimensions, viz., high surface-area-to-volume ratio, leading to a number of extraordinary physical and chemical properties like higher molecular adsorption, large surface tension force, increased chemical and biological activities, greater catalytic effects and very high mechanical strength (Barbieri et al. 2005). Another important key factor of nanomaterials is the quantum size effect that results in their discrete electronic band-structure like those of molecules. Unlike the very high surface-to-volume ratio that also occurs when going from macro to micro dimensions, the quantum effect is only specific to a deep nanoscale dimension of smaller than a few tens of nanometer (Acharya et al. 2009, Tsai et al. 2008). The nanomaterials are thus highly useful for a wide range of nanotechnology fields including nanoelectronics (Tsai et al. 2008), optoelectronics (Kaul et al. 2010), nanophotonics (Botey et al. 2010), nano-electromechanical systems (NEMS) (Kratochvil et al. 2007), bioelectronics (Kim et al. 2010), nanobiotechnology (Euliss et al. 2006), nanochemistry (Stender et al. 2008), biochemistry (Geng et al. 2009), biomedicine (Petrovie et al. 2010), electrochemistry (Chen et al. 2007), nanomechanics (Park et al. 2009) and so on. These lead to a large variety of applications such as quantum-effect lasers/solar cells/transistors (Bernardi et al. 2010), photonic band-gap devices (Jiang et al. 2007), catalyst (Jakhmola et al. 2010, Jiao et al. 2009), photocatalyst (Baikousi et al. 2012, Banerjee et al. 2011), molecular electronic device, surface-enhanced Raman spectroscopy (SERS) (Guo et al. 2011), nano fuel cells (Antolini et al. 2009, Guo et al. 2011), nano-drug delivery systems (Arami et al. 2011, Ochekpe et al. 2009), nanosensors advanced energy storage devices (Simon et al. 2008) and nanoactuators. Among these, sensors are among the fastest-growing applications due to their huge demands in many real-world application fields, such as automobiles, communication, consumer electronics, industrial and biomedical. Sensors can be divided into several classes including mechanical, thermal, optical, magnetic, gas, chemical and biological. Among various kinds of sensors, gas/chemical/ biological sensors can exploit the most benefits from the high surface-to-volume-ratio property of nanomaterials (Basu et al. 2011, Jimenez-Cadena et al. 2007). Gas/chemical/biological sensors generally comprise sensing material that responds to changes of gas/chemical/biological analytes and transducer that converts the changes into electrical signals. The gas sensor may be classified by sensing mechanisms into chemoresistive, surface acoustic wave (SAW), quartz crystal microbalance (QCM), chemiluminescent, optical absorption and dielectric types (Jung et al. 2008). Gas-sensing applications include toxic gases such as nitrogen dioxide, carbon monoxide, sulfur dioxide, ammonia, oxygen and hydrogen sulfide. Similarly, various flammable gases such as hydrogen, methane, acetylene, propane, etc., can also be considered. Gas sensing applications are also very popular among volatile organic compounds (VOCs) such as ethanol, acetone, methanol and propanol (Lu et al. 2009a, Lu et al. 2009b, Shafiei et al. 2010, Arsat et al. 2009, Qin et al. 2008). At the same time, on the basis of sensing platforms, chemical sensors can be classified into electrochemical, ion-sensitive field-effect, chemiluminescent, optical and mass spectroscopic methods (Spencer et al. 2012). It is well known that chemical sensing applications are much wider than gas-sensing ones due to the presence of a large number of liquid-phase chemicals ranging from acids, bases, solvents and inorganic substances to organic analytes (Fowler et al. 2009). Similarly, widely used biosensing platforms include electrochemical, fluorescent, surface plasmon resonance (SPR), QCM and microcantilever (Wang et al. 2005, Wang et al. 2008). Biosensing applications also cover a wider range of biologically relevant materials considering bioanalytes present in living organisms, such as glucose, cholesterol and uric acid, DNAs, RNAs, cells, proteins, organelles and so on (Chopra et al. 2007, Wang et al. 2008). The main and common requirement of these sensors is high sensitivity and specificity. The specific surface area of sensing material is one of the most important factors that dictate the sensitivity as it is directly related to adsorption or reaction rate with target analytes (Song et al. 2011). Gas/chemical/biological sensors developed based on well-established microtechnology are currently used in commercial applications. They provide good sensitivity and reproducibility along with low power consumption. However, their performances are still not satisfactory for many advanced applications that involve the detection of very low concentration analytes. The use of nanomaterials in these sensors will provide a substantial improvement of sensing performances due to several orders of magnitude increase of specific surface area and smaller size (Lin et al. 2009). Well-controlled synthesis and fundamental understanding of properties of nanomaterials are very important for the advancement of nanomaterial-based gas/chemical/biological sensors.

6. Conclusion

This chapter has primarily introduced to the basic knowledge of sensors and sensing technologies. It has also well explained the basic mechanism behind the sensing technology. The basic synthetic strategies of all kinds of nanomaterials, such as top-down, bottom-up and combination, have also been included in this chapter. It has given a brief knowledge of the excellent surface-to-volume ratio of nanomaterials and their effect on potential sensing property.

References

Acharya, S., D.D. Sarma, Y. Golan. 2009. Shape-dependent confinement in ultra small zero-, one-, and two-dimensional pbs nanostructures. *J. Am. Chem. Soc.* 131, 11282–11283.

Antolini, E. 2009. Carbon supports for low-temperature fuel cell catalysts. *Appl. Catal. Environ.* 88, 1–24.

Arami, H., Z. Stephen, O. Veiseh. 2011. Chitosan-coated iron oxide nanoparticles for molecular imaging and drug delivery. *Adv. Polym. Sci.* 243, 169–184.

Arsat, R., M. Breedon, M. Shafiei. 2009. Graphene-like nano-sheets for surface acoustic wave gas sensor applications. *Chem. Phys. Lett.* 467, 344–347.

Baikousi, M., A.B. Bourlinos, A. Douvalis. 2012. Synthesis and characterization of g-Fe$_2$O$_3$/carbon hybrids and their application in removal of hexavalent chromium ions from aqueous solutions. *Langmuir* 28, 3918–3930.

Banerjee, A.N. 2011. The design, fabrication, and photocatalytic utility of nanostructured semiconductors: focus on TiO$_2$-based nanostructures. *Nanotechnol. Sci. Appl.* 4, 35–65.

Bang, J., J. Bae, P. Lwenhielm. 2007. Facile routes to patterned surface neutralization layers for block copolymer lithography. *Adv. Mater.* 19, 4552–4557.

Banica, F.G. 2012. Potentiometric sensors. pp. 165–216. *In*: Chemical Sensors and Biosensors: Fundamentals and Applications. Chichester, UK, John Wiley & Sons.

Bao, C., F. Tian, G. Estrada. 2010. Improved visualization of internalized carbon nanotubes by maximising cell spreading on nanostructured substrates. *Nano Biomed. Eng.* 2, 201–207.

Barbero, C.A., D.F. Acevedo, E. Yslas. 2010. Synthesis, properties and applications of conducting polymer nano-objects. *Mol. Cryst. Liq. Cryst.* 521, 214–228.

Barbieri, O., M. Hahn, A. Herzog. 2005. Capacitance limits of high surface area activated carbons for double layer capacitors. *Carbon* 43, 1303–1310.

Basu, S., P.K. Basu. 2011. Nanomaterials and chemical sensors. *Sens. Transducers* 134, 1–31.

Bernardi, M., M. Giulianini, J.C. Grossman. 2010. Self-assembly and its impact on interfacial charge transfer in carbon nanotube/p3ht solar cells. *ACS Nano.* 4, 6599–6606.

Botey, M., J. Martorell, G. Lozano. 2010. Anomalous group velocity at the high energy range of real 3d photonic nanostructures. *In*: Proceedings of SPIE—The International Society for Optical Engineering.

Cadena, R. 2013. Electromechanical systems. *In*: Automated Lighting: The Art and Science of Moving Light in Theatre, Live Performance, Broadcast and Entertainment, Burlington, MA, Elsevier.

CAPGO. Introduction to Semiconductor Temperature Sensors, http://www.capgo. com/Resources/Temperature/ Semiconductor/Semi.html, 2010.

Chan, W.C.W., D.J. Maxwell, X. Gao. 2002. Luminescent quantum dots for multiplexed biological detection and imaging. *Curr. Opin. Biotechnol.* 13, 40–46.

Chen, D., Y. Gao, G. Wang. 2007. Surface tailoring for controlled photoelectrochemical properties: effect of patterned tio2 microarrays. *J. Phys. Chem. C* 111, 13163–13169.

Chen, J., F. Cheng. 2009. Combination of light weight elements and nanostructured materials for batteries. *Acc. Chem. Res.* 42, 713–723.

Chen, J.S., L.A. Archer, W. Lou. 2011. SnO_2 hollow structures and TiO_2 nanosheets for lithium-ion batteries. *J. Mater. Chem.* 21, 9912–9924.

Chen, R.J., S. Bangsaruntip, K.A. Drouvalakis. 2003. Noncovalent functionalization of carbon nanotubes for highly specific electronic biosensors. *Proc. Natl. Acad. Sci. USA* 100(9), 4984–4989.

Chen, X., S.S. Mao. 2007. Titanium dioxide nanomaterials: synthesis, properties, modifications and applications. *Chem. Rev.* 107, 2891–2959.

Chen, Y., Z. Xu, M.R. Gartia. 2010. Ultrahigh throughput silicon nanomanufacturing by simultaneous reactive ion synthesis and etching. *ACS Nano.* 5, 8002–8012.

Chopra, N., V.G. Gavalas, B.J. Hinds. 2007. Functional one-dimensional nanomaterials: applications in nanoscale biosensors. *Anal. Lett.* 40, 2067–2096.

Ciesielski, A., C.A. Palma, M. Bonini. 2010. Towards supramolecular engineering of functional nanomaterials: pre-programming multi-component 2d self-assembly at solid-liquid interfaces. *Adv. Mater.* 22, 3506–3520.

Colaianni, L., S.C. Kung, D. Taggart. 2009. Gold nanowires: deposition, characterization and application to the mass spectrometry detection of low-molecular weight analytes. pp. 20–24. *In*: 3rd international workshop on advances in sensors and interfaces (IWASI 2009). June 25–26, 2009, Trani (Bari), Italy.

Colvin, V.L. 2003. The potential environmental impact of engineered nanomaterials. *Nat. Biotechnol.* 21, 1166–1170.

Euliss, L.E., J.A. DuPont, S. Gratton. 2006. Imparting size, shape, and composition control of materials for nanomedicine. *Chem. Soc. Rev.* 35, 1095–1104.

Fang, X., L. Zhang. 2006. One-dimensional (1d) ZnS nanomaterials and nanostructures. *J. Mater. Sci. Technol.* 22, 721–736.

Fink, J.K. 2012. Mechanical sensors. pp. 131–138. *In*: Polymeric Sensors and Actuators. Hoboken, Massachusetts, Wiley-Scrivener.

Forey, C., C. Mellot-Draznieks, C. Serre. 2005. Chemistry: a chromium terephthalate based solid with unusually large pore volumes and surface area. *Science* 309, 2040–2042.

Fowler, J.D., M.J. Allen, V.C. Tung. 2009. Practical chemical sensors from chemically derived graphene. *ACS Nano.* 3, 301–306.

Fraden, Jacob. 2010. Temperature sensors. pp. 519–567. *In*: Handbook of Modern Sensors. Springer New York.

Gasparotto, A., D. Barreca, C. MacCato. 2012. Manufacturing of inorganic nanomaterials: concepts and perspectives. *Nanoscale* 4(9), 2813–2825.

Geng, L., P. Jiang, J. Xu. 2009. Applications of nanotechnology in capillary electrophoresis and microfluidic chip electrophoresis for biomolecular separations. *Prog. Chem.* 21, 1905–1921.

George, J., K.V. Ramana, A.S. Bawa. 2011. Bacterial cellulose nanocrystals exhibiting high thermal stability and their polymer nanocomposites. *Int. J. Biol. Macromol.* 48, 50–57.

Guo, S., E. Wang. 2011. Functional micro/nanostructures: simple synthesis and application in sensors, fuel cells, and gene delivery. *Acc. Chem. Res.* 44, 491–500.

Guth, U., F. Gerlach, M. Decker, W. Oelsner, W. Vonau. 2009. Solid-state reference electrodes for potentiometric sensors. *J. Solid State Electr.* 13, 27–39.

Gyorki, John R. 2009. Designing with Semiconductor Temperature Sensors, Last Update: 2009, http://www.sensortips.com/temperature/designing-with-semiconductor-temperature-sensors/.

Heberto, G.P., J.L. Gonzalez-Vidal, G.A. Torres, J.R. Baez, A. Maldonado, M. Olvera, D.R. Acosta, M.A. Alejo, L. Castaneda. 2013. Chromium and ruthenium-doped zinc oxide thin films for propane sensing applications. *Sensors* 13, 3432–3444.

Hillhouse, H.W. 2011. Development of double-gyroid nanowire arrays for photovoltaics. *In*: Proceedings of the 2011 AIChE annual meeting. October 16–21, 2011, Minneapolis, MN.

Ivanovskii, A.L. 2003. Fullerenes and related nanoparticles encapsulated in nanotubes: synthesis, properties, and design of new hybrid nanostructures. *Russ. J. Inorg. Chem.* 48, 846–860.

Jakhmola, A., R. Bhandari, D.B. Pacardo. 2010. Peptide template effects for the synthesis and catalytic application of pd nanoparticle networks. *J. Mater. Chem.* 20, 1522–1531.

Jeong, S., H.C. Shim, S. Kim. 2009. Efficient electron transfer in functional assemblies of pyridine-modified nQDs on SWNTs. *ACS Nano.* 4, 324–330.

Jiang, P., C.H. Sun, N.C. Linn. 2007. Self-assembled photonic crystals and templated nanomaterials. *Curr. Nanosci.* 3, 296–305.

Jiao, H. 2009. Recent developments and applications of iron oxide nanomaterials. *FenmoYejinCailiaoKexueyuGongcheng/Mater. Sci. Eng. Powder Metallurgy* 14, 131–137.

Jimenez-Cadena, G., J. Riu, F.X. Rius. 2007. Gas sensors based on nanostructured materials. *Analyst* 132, 1083–1099.

Jung, I., D. Dikin, S. Park. 2008. Effect of water vapor on electrical properties of individual reduced graphene oxide sheets. *J. Phys. Chem. C* 112, 20264–20268.

Katz, E., I. Willner. 2004. Integrated nanoparticle-biomolecule hybrid systems: synthesis, properties, and applications. *Angew. Chem. Int. Ed.* 43, 6042–6108.

Kaul, A.B., K. Megerian, L. Bagge. 2010. Carbon-based nanodevices for electronic and optical applications. pp. 304–307. *In*: Nanotechnology 2010: Electronics, Devices, Fabrication, MEMS, Fluidics and Computational—Technical Proceedings of the 2010 NSTI Nanotechnology Conference and Expo, NSTI-Nanotech.

Khanna, Vinod Kumar. 2012. Nanosensors: Physical, Chemical, and Biological. Boca Raton: CRC Press.

Kim, J., B.C. Kim, D. Lopez-Ferrer. 2010. Nanobiocatalysis for protein digestion in proteomic analysis. *Proteomics* 10, 687–699.

Knott, Barry. 2010. Semiconductor Technology—Personal Breathalyzer, Last Update: January 10th 2010, http://ezinearticles.com/?Semiconductor-Technology—Personal-Breathalyzers&id=3511961.

Kratochvil, B.E., L. Dong, L. Zhang. 2007. Automatic nanorobotic characterization of anomalously rolled-up sige/si helical nanobelts through vision-based force measurement. pp. 57–62. *In*: Proceedings of the 3rd IEEE international conference on automation science and engineering (IEEE CASE 2007). September 22–25, 2007, Scottsdale, AZ, USA.

Lam, C.W., J.T. James, R. McCluskey. 2004. Pulmonary toxicity of single-wall carbon nanotubes in mice 7 and 90 days after intratracheal instillation. *Toxicol. Sci.* 77, 126–134.

Lin, Y.M., K.A. Jenkins, V.G. Alberto. 2009. Operation of graphene transistors at giqahertz frequencies. *Nano Lett.* 9, 422–426.

Liu, L., K. Busuttil, S. Zhang. 2011. The role of self-assembling polypeptides in building nanomaterials. *Phys. Chem. Chem. Phys.* 13, 17435–17444.

Lo, P.K., P. Karam, F.A. Aldaye. 2010. Loading and selective release of cargo in DNA nanotubes with longitudinal variation. *Nat. Chem.* 2, 319–328.

Lu, G., L.E. Ocola, J. Chen. 2009a. Gas detection using low-temperature reduced graphene oxide sheets. *Appl. Phys. Lett.* 94, 8.

Lu, G., L.E. Ocola, J. Chen. 2009b. Reduced graphene oxide for room-temperature gas sensors. *Nanotechnology* 20, 445502–445510.

Lu, J., Y. Dongning, L. Jie. 2008. Three dimensional single-walled carbon nanotubes. *Nano Lett.* 8, 3325–3329.

Madsen, M., K. Takei, R. Kapadia. 2011. Nanoscale semiconductor "X" on substrate "Y"—Processes, devices, and applications. *Adv. Mater.* 23, 3115–3127.

Martin, C.R. 1994. Nanomaterials: a membrane-based synthetic approach. *Science* 266, 1961–1966.

Martin, C.R. 1996. Membrane-based synthesis of nanomaterials. *Chem. Mater.* 8, 1739–1746.

McGrath, M.J., C.N. Scanaill. 2013. Sensor Technologies, Healthcare, Wellness and Environmental Applications. Apress Open.

Nel, A., T. Xia, L. Mudler. 2006. Toxic potential of materials at the nano level. *Science* 311, 622–627.

Nihal, K., B.H. Sudantha. 2008. An environmental air pollution monitoring system based on the IEEE 1451 standard for low cost requirements. *IEEE Sens. J.* 8, 415–422.

Oberdarster, G., E. Oberdarster, J. Oberdarster. 2005. Nanotoxicology: an emerging discipline evolving from studies of ultrafine particles. *Environ. Health. Perspect.* 113, 823–839.

Ochekpe, N.A., P.O. Olorunfemi, N.C. Ngwuluka. 2009. Nanotechnology and drug delivery. Part 1: background and applications. *Trop. J. Pharm. Res.* 8, 265–274.

Okamoto, H., Y. Sugiyama, H. Nakano. 2011. Synthesis and modification of silicon nanosheets and other silicon nanomaterials. *Chem. Eur. J.* 17, 9864–9887.

Park, H.S., W. Cai, H.D. Espinosa. 2009. Mechanics of crystalline nanowires. *MRS Bull.* 34, 178–183.

Pashchanka, M., R.C. Hoffmann, A. Gurlo. 2010. Molecular based, chimiedouce approach to 0d and 1d indium oxide nanostructures. Evaluation of their sensing properties towards co and h 2. *J. Mater. Chem.* 20, 8311–8319.

Patranabis, D. 2004. Sensors and Transducers. 2nd ed. New Delhi: PHI Learning Pvt Ltd.

Patzke, G.R., F. Krumeich, R. Nesper. 2002. Oxidic nanotubes and nanorods—anisotropic modules for a future nanotechnology. *Angew. Chem. Int. Ed.* 41, 2446–2461.

Petrovie, Z.L., M. Radmilovic-Radenovic, P. Maguire. 2010. Application of nonequilibrium plasmas in top-down and bottom-up nanotechnologies and biomedicine. pp. 29–36. *In*: Proceedings of the 2010 27th international conference on microelectronics (MIEL 2010). May 16–19, 2010, Aleksan Nis, Serbia.

Pietryga, J.M., K.K. Zhuravlev, M. Whitehead. 2008. Evidence for barrierless auger recombination in PbSe nanocrystals: a pressure-dependent study of transient optical absorption. *Phys. Rev. Lett.* 101, 217401.

Pokropivny, V.V., V.V. Skorokhod. 2008. New dimensionality classifications of nanostructures. *Physica E* 40, 2521–2525.

Qin, L., J. Xu, X. Dong. 2008. The template-free synthesis of square-shaped SnO_2 nanowires: the temperature effect and acetone gas sensors. *Nanotechnology* 19, 18.

Sayle, T.X.T., R.R. Maphanga, P.E. Ngoepe. 2009. Predicting the electrochemical properties of MnO_2 nanomaterials used in rechargeable Li batteries: simulating nanostructure at the atomistic level. *J. Am. Chem. Soc.* 131, 6161–6173.

Seok, H., T.H. Park. 2011. Integration of biomolecules and nanomaterials: Towards highly selective and sensitive biosensors. *Biotechnol. J.* 6, 1310–1316.

Shafiei, M., P.G. Spizzirri, R. Arsat. 2010. Platinum/graphene nanosheet/SiC contacts and their application for hydrogen gas sensing. *J. Phys. Chem. C* 114, 13796–13801.

Simon, P., Y. Gogotsi. 2008. Materials for electrochemical capacitors. *Nat. Mater.* 7, 845–854.

Sondi, I., B. Salopek-Sondi. 2004. Silver nanoparticles as antimicrobial agent: a case study on *E. Coli* as a model for gram-negative bacteria. *J. Colloid. Interface. Sci.* 275, 177–182.

Song, H.S., W.J. Zhang, C. Cheng. 2011. Controllable fabrication of three-dimensional radial ZnO nanowire/silicon microrod hybrid architectures. *Cryst. Growth Des.* 11, 147–153.

Spencer, M.J.S. 2012. Gas sensing applications of 1d-nanostructured zinc oxide: insights from density functional theory calculations. *Prog. Mater. Sci.* 57, 437–486.

Stender, C.L., P. Sekar, T.W. Odom. 2008. Solid-state chemistry on a surface and in a beaker: unconventional routes to transition-metal chalcogenide nanomaterials. *J. Solid State Chem.* 181, 1621–1627.

Sun, E.Y., L. Josephson, R. Weissleder. 2006. "Clickable" nanoparticles for targeted imaging. *Mol. Imaging* 5, 122–128.

Tsai, C., R.J. Tseng, Y. Yang. 2008. Quantum dot functionalized one dimensional virus templates for nanoelectronics. *J. Nanoelectron. Optoelectron.* 3, 133–136.

Wang, J. 2005. Carbon-nanotube based electrochemical biosensors: a review. *Electroanalysis* 17, 7–14.

Wang, J. 2008. Electrochemical glucose biosensors. *Chem. Rev.* 108, 814–825.

Wang, Y., Z. Li, J. Wang. 2011. Graphene and graphene oxide: biofunctionalization and applications in biotechnology. *Trends Biotechnol.* 29, 205–212.

Wetchakun, K., T. Samerjai, N. Tamaekong, C. Liewhiran, C. Siriwong, V. Kruefu, A. Wisitsoraat, A. Tuantranont, S. Phanichphant. 2011. Semiconducting metal oxides as sensors for environmentally hazardous gases. *Sens. Actuators B Chem.* 160, 580–591.

Yi, D.K., J.H. Lee, J.A. Rogers. 2009. Two-dimensional nanohybridization of gold nanorods and polystyrene colloids. *Appl. Phys. Lett.* 94, 084104.

Zhang, S. 2003. Fabrication of novel biomaterials through molecular self-assembly. *Nat. Biotechnol.* 21, 1171–1178.

Zhao, Y.S., H. Fu, A. Peng. 2008. Low-dimensional nanomaterials based on small organic molecules: Preparation and optoelectronic properties. *Adv. Mater.* 20, 2859–2876.

Zhao, Y.S., H. Fu, A. Peng. 2009. Construction and optoelectronic properties of organic one-dimensional nanostructures. *Acc. Chem. Res.* 43, 409–418.

Chapter **2**

Metal Oxides Nanomaterials in Gas Sensors

1. Introduction

The continuous increase in environmental pollution through factory wastes, exhaust gases from motor vehicles and other urbanization issues have continually made effect in the release of toxic, explosive, flammable and carcinogenic gaseous products in the environment of all developed and developing countries. A clean air supply is a primary requirement for human health. The human nose can serve as a highly advanced sensing system, although it can only detect the bad odor of foul gases but cannot sense inhalation of odorless toxic gaseous exhausts. Therefore, a high rate of gas emissions can make a tremendous negative effect on human/animal health, which in consequence cause a direct adverse effect on the environment and natural resources. The process to detect the concentration of toxic gases introduced the development of gas sensor materials. The gas sensors can detect combustible, toxic and explosive gaseous materials if concentration crosses individual optimum value. They can notify through an alarm system (e.g., signal or sound) attached to the portable devices. The 4S parameters (stability, sensitivity, selectivity and speed) are of the utmost importance in the process of sensing. Besides them, response and recovery time with power consumption are also playing key factors. The sensors usually record a change in the physical conditions of the target gas. At the first step, there occurs a collision between target gas with surface atoms, which transforms into adsorption/desorption of gas on the sensing material surface at a particular operating temperature. The concentration of the target gas will further generate a signal (Thomas et al. 2015, Liu et al. 2012, Nazemi et al. 2019).

In recent years, nanomaterials have drawn huge attention in every sector of science and engineering due to their very high surface-to-volume ratio and significant geometrical structure. Nano-sized materials have been extensively used as potential gas sensors because of their low cost, easy production, compact size and simple measuring electronics (Tomchenko et al. 2003). The special morphology and structure with effective particle size, surface area and porosity of nanomaterials influence the performance of these sensors over bulk materials or dense films through their highly sensitive properties. There are many metal oxides that have already been reported as a potential candidate in the detection of combustible gases through their conductive measurements, such as Cr_2O_3, Mn_2O_3, Co_3O_4, NiO, CuO, SrO, In_2O_3, WO_3, TiO_2, V_2O_3, Fe_2O_3, GeO_2, Nb_2O_5, MoO_3, Ta_2O_5, La_2O_3, CeO_2 and Nd_2O_3 (Kanazawa et al. 2001). The selection of metal oxides primarily depends on their electronic structure. On basis of that, those have been categorized into two groups: transition (NiO, Fe_2O_3, etc.) and non-transition metal oxides (Al_2O_3). These non-transition metal oxides are also divided into two groups, viz., pre-transition (Al_2O_3) and post-transition metal oxides (ZnO, SnO_2, etc.). Due to the presence of long band gaps, pre-transition metal oxides have acted inert. As they can neither form electrons nor holes very easily, they rarely consider as a gas sensing material for their poor electrical conductivity nature. On the other hand, the energy difference between a cation with d_n configuration and either a d_{n+1} or d_{n-1} configuration is often very small (Henrich et al. 1994).

2. Mechanism Behind the Gas Sensing in Metal Oxides

To date, metal oxides have been potentially applied in gas sensing with appropriate designing and fabrication, which induced the performance many fold. However, the fundamental mechanisms that cause a gas response are still controversial, and it is essentially responsible for a change in conductivity that allows trapping of electrons at adsorbed molecules and band-bending induced by these charged molecules. Herein, a brief introduction to the sensing mechanism of n-type metal oxides in the air is given based on the example of SnO_2. Typically, oxygen gases are adsorbed on the surface of the SnO_2 sensing material in the air. The adsorbed oxygen species can capture electrons from the inner of the SnO_2 film. The negative charge trapped in these oxygen species causes a depletion layer and thus a reduced conductivity. When the sensor is exposed to reducing gases, the electrons trapped by the oxygen adsorbate will return to the SnO_2 film, leading to a decrease in the potential barrier height and thus an increase in conductivity. There are different oxygen species including molecular (O_2^-) and atomic (O^-, O^{2-}) ions on the surface depending on working temperature. Generally, below 150°C the molecular form dominates while above this temperature the atomic species are found (Tricoli et al.

2010, Barsan et al. 2001). The overall surface stoichiometry has a decisive influence on the surface conductivity of the metal oxides. Oxygen vacancies act as donors, increasing the surface conductivity, whereas adsorbed oxygen ions act as surface acceptors, binding elections and diminishing the surface conductivity. Figure 2.1 shows the energy diagram of various oxygen species in the gas phase, adsorbed at the surface and bound within the lattice of SnO_2 (Liu et al. 2007, Kohl et al. 1989). On SnO_2 films, the reaction

$$O_{2\,ads}^- + e^- = 2O_{ads}^-$$

takes place as the temperature increases. The desorption temperatures from the SnO_2 surface are around 550°C for O_{ads}^- ions and around 150°C for $O_{2\,ads}^-$ ions. At constant oxygen coverage, the transition causes an increase in surface charge density with corresponding variations of band-bending and surface conductivity. From conductance measurements, it is concluded that the transition takes place slowly. Therefore, a rapid temperature change on the part of the sensors is usually followed by a gradual and continuous change in the conductance. The oxygen coverage adjusts to a new equilibrium and the adsorbed oxygen is converted into another species which may be used in measurement method of dynamic modulated temperature as reported previously (Huang et al. 2003a, Huang et al. 2004b, Huang et al. 2004b, Huang et al. 2004c, Huang et al. 2003b, Huang et al. 2005, Sun et al. 2004).

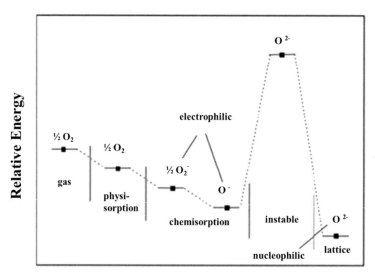

Figure 2.1: Energy diagram for various oxygen species in the gas phase adsorbed at the surface and bound within the lattice of SnO_2. Reprinted with permission from Liu et al. 2007. Copyright (2007) Nova Science Publishers.

3. Effect of Nano-particles in Sensing

There are many research groups who have actively reported the "small size effect" of metal oxides (Khare et al. 2005, Aswal et al. 2002, Kaur et al. 2005, Bhandarkar et al. 2006, Aswal et al. 2005, Yamazoe et al. 2009a, Yamazoe et al. 2009b). It is clearly seen in Figure 2.2, a sensor can be taken into account as they are composed of partially sintered crystallites which can be connected to their neighbors by necks. Those interconnected grains can produce larger aggregates that are connected to their neighbors via grain boundaries (Rothschild et al. 2004). On the surface of the grains, adsorbed O_2 extract e⁻s from the conduction band and capture the electrons on the surface as ionic forms; they can further produce a band-bending and an electron depleted region, which is termed as the space-charge layer. In the case when the particle size of the sensing film is closer to or lesser than double the thickness of the space-charge layer, the sensitivity of the sensor will enhance extraordinarily. Xu et al. reported the phenomena by using a semiquantitative model (Xu et al. 1991). Three different cases can be distinguished according to the relationship between the particle size (D) and the width of the space-charge layer (L) that is produced around the surface of the crystallites as chemisorbed ions and the size of L is about 3 nm in the case of pure SnO_2 material (Ogawa et al. 1982, Yin et al. 2009, Nguyen et al. 2008, Zhang et al. 2009, Meng et al. 2009). When the condition is like $D \gg 2L$, the conductivity of the whole structure actually controls the inner mobile charge carriers and the electrical conductivity depends exponentially on the barrier height. It is not so sensitive to the charges received from surface reactions. On the other hand, when $D \geq 2L$, the space-charge layer region around each neck yields a constricted conduction channel within each aggregated structure. In consequence, the conductivity not only is controlled by the particle boundaries barriers but also by the cross-section area of those channels, so it becomes very sensitive to reaction charges. Thus, the particles are sensitive to ambient gas composition. Similarly, when the condition is maintained as $D < 2L$, the space-charge layer region dominates the whole particle, and the crystallites are almost fully depleted of mobile charge carriers. The energy bands are leading into nearly flat throughout the whole structure of the inter-connected grains and no significant barriers are developed for inter-crystallite charge transport and then the conductivity value becomes essentially controlled by the inter-crystallite conductivity. Few charges received from surface reactions will result in a large variation of conductivity in the whole structure, therefore the crystalline SnO_2 becomes sensitive in considerable amount to ambient gas molecules when its particle size is small enough. Following Xu's model, many new sensing materials are reported with high gas sensing properties (Li et al.

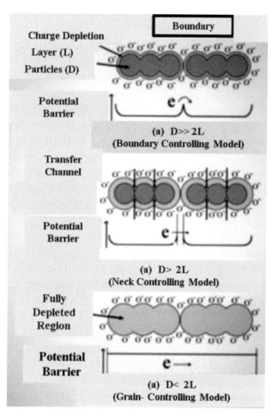

Figure 2.2: Schematic model of the effect of the crystallite size on the sensitivity of metal-oxide gas sensors: (a) D >> 2L; (b) D ≥ 2L; (c) D < 2L (reproduced with permission from Sun et al. 2012) (Copyright MDPI).

2011, Zhang et al. 2011, Yu et al. 2011). Typically, the nano-composite of SnO_2 and multiwall carbon nanotube (MWCNT) can be exploited to detect persistent organic pollutants (POPs) which constitutes table chemical properties and are ordinarily difficult to be detected by metal oxides (Meng et al. 2010). The development of materials with different sizes and porosity ranges in the nano-meter range plays a key role in respect to the technological importance in a wider range of sensing applications. In a work, it is reported that the ultrasensitive detection of aldrin and dichlorodiphenyltrichloroethane (DDT) has been carried out by applying the nanocomposite of small SnO_2 particles and MWCNTs. In this case, the nanocomposite generates a very attractive improved sensitivity in comparison with the activity of a conventional SnO_2 sensor. A sharp response of low limiting concentration of about 1 ng was observed in both aldrin and DDT, suggesting potential applications as a new analytical approach. On the other hand, the SnO_2/MWCNT nanocomposite prepared by a wet chemical method may control the particle size of SnO_2 under 10 nm and produce highly porous three dimensional (3D) structures. Among the highly porous 3D structures, MWCNTs can be considered as the framework and the SnO_2 particles are uniformly packed on them, which may increase the ability of gas diffusion performance into and out of the sensing film. The high sensitivity can also be affirmed due to an effect of the p-n junction formed between p-type carbon nanotubes and n-type SnO_2 nanoparticles. The investigation results make SnO_2/MWCNT nanocomposites attractive for the purpose of POPs' detection.

4. Metal-Oxide Hollow Sphere Synthetic Strategies

Hard templating methods were widely employed in the earlier stage for the production of multi-shelled hollow structures due to their simple and straightforward concepts. In 1998, Caruso et al. introduced the colloidal templated electrostatic layer-by-layer (LBL) self-assembled structure of silica nanoparticles with polymer multi-layers. These nanoparticles electrostatically self-assemble onto the linear cationic polymer poly(diallyl dimethyl ammonium chloride) (PDADMAC) (Caruso et al. 1998). In this method, the core template structure was removed through the calcination process and/or exposure to the solvent. The thickness of the wall and overall shape can be controlled with a number of layers of SiO_2–PDADMAC present in deposition cycles. In a typical multi-step process, the negatively charged PS latex particles were used as core material, on which three-layered smooth and positively charged polymer film was deposited. This multilayer film in the actual sense induced the subsequent adsorption of SiO_2. Figure 2.3 illustrates the synthetic process of typical inorganic and hybrid hollow spheres.

In the next step, the shell-by-shell assembly combined with colloidal templating to form multi-shelled structures had been adopted. During this process, the sequential coating of precursor materials developed onto the surface of a hard template and another material as an interlayer. By repeating the same coating process, multiple shells had been formed. Subsequently, removing the template and interlayer had resulted in the multi-layered hollow structure. A similar synthetic method had been adopted by Lou et al. in 2007 to prepare double-shelled SnO_2 hollow colloids using shell-by-shell hydrothermal deposition onto silica nano-templates (Lou et al. 2007). In 2011, Huang et al. had reported the preparation of double- and triple-shelled silica nanoparticles with a coating of mesoporous silica layer, which further resulted in the formation of PS@silica nanoparticles via a CTAB surfactant-assisted process followed by self-templating etching technique (Huang et al. 2011). The shell-to-shell distance had been

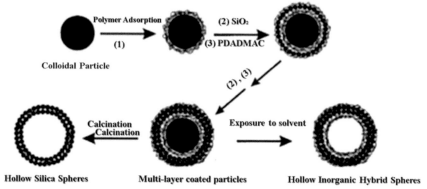

Figure 2.3: Illustration of procedures for preparing inorganic and hybrid hollow spheres. The scheme is shown for PS latex particles (reproduced with permission from Caruso et al. 1998).

controlled by varying the etching period, without a change in the multi-shelled size or interior core diameter.

Wong et al. reported the synthesis of triple-shelled hollow Au@SiO$_2$ nanospheres using the typical Stöber Method. They had provided specific conditions to etch or harden the silica shells, resulting in the formation of inhomogeneous core-shell nanostructures. In this work, methods were introduced to harden the soft silica shells in which the outer shell proved to be more robust than the inner layer with selective etching by hot water (Wong et al. 2011). Yang et al. followed the facile layer-by-layer (LBL) self-assembly method for controllable preparation of multi-shelled NiO hollow nanospheres by calcination of Ni(OH)$_2$/C precursors at different stages with the repeated dispersion with urea and NiCl$_2$ (Yang et al. 2014). In this typical process, the crystal nucleation occurred after the formation of Ni(OH)$_2$ mesocrystals from supersaturated solution, resulting in the declining concentration of Ni^{2+} ions in solution. Therefore, a thin layer of Ni(OH)$_2$ nanoparticles produced above and under the carbon sphere surfaces. In consequence, the Ni(OH)$_2$ nanoparticles inside carbon spheres could not assemble into nanoflakes because of the confinement within carbon spheres. Figure 2.4 shows the synthetic scheme with layer-by-layer (LBL) method (Yang et al. 2014).

Wu et al. in 2009 described a new and convenient route to prepare double-layered hollow spherical structures using solid templates by programming the reaction-temperature methods (Yu et al. 2007). They initiated the process with the solid templates of V(OH)$_2$NH$_2$, and they made a relationship between the rate and temperature of the hollowing process using the Van't Hoff rule and the Arrhenius activation-energy equation. They finally concluded that chemical reaction controlled the hollowing process over the diffusion. Yang et al. had also proposed a potential

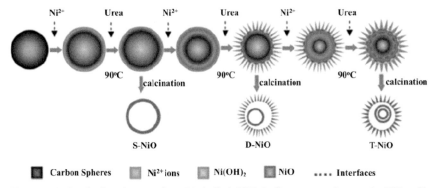

Figure 2.4: Synthetic scheme of multi-shelled NiO hollow nanospheres via LBL self-assembly (reproduced with permission from Yang et al. 2014).

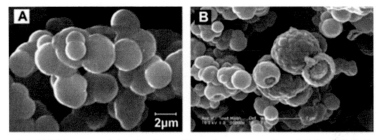

Figure 2.5: SEM images of (a) the tin oxide-carbon composites, (b) the HS-SnO$_2$ after calcination (reproduced with permission from Yang et al. 2007).

synthetic method to produce multi-shelled SnO$_2$ hollow microspheres using a chemically induced hydrothermal technique (Yang et al. 2007). The multi-shelled SnO$_2$-carbon composites were formed using condensation polymerization and carbonization of sucrose with the hydrolysis of SnCl$_4$ under solvothermal conditions, followed by the calcination to remove the core carbon. However, they had not mentioned the key factor to adjust the shell number. Therefore, this method had not been taken into account further by any other research groups.

Yu's research group had described the synthetic strategies of vesicular silica and functionalized vesicular silica formed with various surfactants (Yu et al. 2007, Wang et al. 2007, Wang et al. 2006). They had used P123 as the shape controlling chemical agent at a mild pH range. The application of Pluronic P85 had also been reported as the single template, multi-layered organo-siliceous vesicles with sponge-like walls, and controllable shell numbers were also synthesized by adjusting the pH value of solutions. Another work reported on the synthesis of nano-sized multi-layered silica vesicles using a dual template method by taking CTAB, FC-4 and

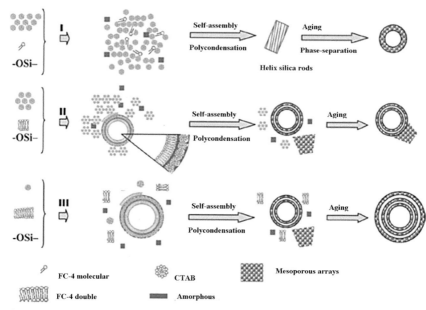

Figure 2.6: Possible mechanism of the formation of silica vesicles (reproduced with permission from Gu et al. 2009).

tetraethyl orthosilicate (TEOS) (Gu et al. 2009). Figure 2.6 presented the possible pathways to form silica vesicles.

However, these traditional synthetic approaches suffer from serious limitations due to their complex and tedious methodology. Additionally, all these processes lack to establish a compatible relationship between the template and the target material. At the same time, the uniform coating could not be formed using these methods. By avoiding these constraints to synthesize multi-shelled hollow spheres, Dan Wang and co-workers had established an outstanding approach using sequential templates. They had demonstrated a remarkable new technique in which traditional tedious multi-step layer-by-layer (LBL) methods can easily be avoided. Their synthetic technique had also proved that without creating a core-shell structure with multiple shells and solid templating material, a multi-layered hollow sphere can still be formed. In 2009, their group introduced a general method to form spinel ferrites hollow spheres (MFe_2O_4), where M = Zn, Co, Ni, Cd (Li et al. 2009). In this work, saccharide microspheres were taken into account as a template, which adsorbed the metal ions and finally hollow structures formed due to gradual removal of the templates by application of heating. It was found that the core size and shell thickness got increased with a higher concentration of metal salt. At the same time, more concentration eliminated the formation of impure phases of ferrites.

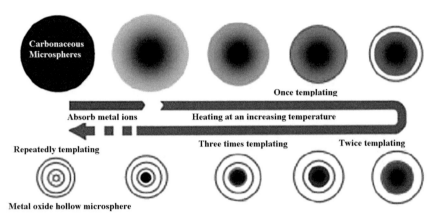

Figure 2.7: Illustration of the sequential templating approach to multiple-shell hollow metal oxide microsphere synthesis (reproduced with permission from Lai et al. 2011).

They have termed this synthetic method as a sequential templating approach (STA). Figure 2.7 shows the illustration of STA to synthesize multi-shelled hollow metal oxides microspheres (Lai et al. 2011).

The same research group had successfully synthesized multi-shelled metal oxide hollow sphere, viz., Fe_2O_3, Co_3O_4, NiO, CuO and ZnO using STA method (Lai et al. 2011). In this typical process, the exterior of the carbonaceous microspheres got easily combusted into gases and released in comparison to the interior due to a temperature gradient along the radial direction, resulting in the reduction of their size. The metal ions concentration gradually increased onto the exterior of the spheres until a certain threshold value could be achieved. During this tenure, metal ions got converted into metal oxides, producing a rigid shell structure. After the threshold heating point, the outer metal oxide shell got detached from the inner template particles. The metal ions formed at the inner template particles could be allowed to develop more metal oxide shells through a similar templating technique. Now the main challenge in this method to define the threshold value. The authors reported that the metal ions achieved this threshold value within a shorter period if the higher level of original concentration had been taken. In this way, one can easily create multiple target shells with different sizes from the same template just by varying the concentration and radial distribution of metal ions within the carbonaceous template. The resultant products can be different from the variation of metal salt concentration, solvent type, pH value, heating temperature and rate and atmosphere type.

The multi-shelled hollow structures were synthesized using a soft-templating method with a relatively flexible multi-layer structure by using supramolecular micelles and polymer vesicles. Although, the multi-

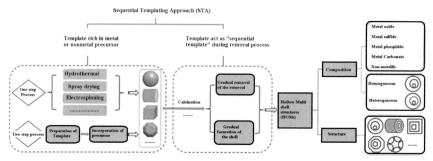

Figure 2.8: Illustration of the sequential templating approach for the synthesis of HoMSs (reproduced with permission from Mao et al. 2018).

shelled structure formed using the soft templating process are metastable thermodynamically and easily gets raptured with the variation in various parameters resulting from shell growth, such as the pH value, temperature, solvent, ionic strength, the concentration of organic templates and inorganic additives. In 2007, Xu et al. reported the formation of multi-shelled Cu_2O hollow spheres by self-assembling of surfactant molecules in an aqueous solution, which further lead to the formation of micelles and closed aggregated vesicles (Xu et al. 2007). They had processed the method with the assistance of CTAB vesicles and multi-lamellar vesicles. They had controlled the number of shells and structures by adjusting the CTAB concentrations. In recent times, template-free methods have become very popular. In this method, the self-assembly of target materials is taken into account without using any templates. One recent work was also mentioned to form silica multi-layered spheres by Soltani et al. (Soltani et al. 2020). In 2008, Zhao et al. had proposed a self-templating synthetic strategy with an application of azithromycin microspheres (Zhao et al. 2008). These azithromycin spheres were prepared in an ethanol-water mixture by applying significantly different solubility in water and ethanol. In this process azithromycin started precipitating with the concentration gradient, resulting in the formation of the first shell. In the next step, azithromycin molecules had continued to diffuse toward the outer side and another shell formed. With the repetition of the same process, multi-shelled spheres had been generated. In this process, the shell number could be easily adjusted by changing the starting concentration of the azithromycin ethanol solution. In very recent time, Zuo et al. reported a synthetic method to form carbon-coated multi-shelled $NiCo_2S_4$ by template-free method (Zuo et al. 2020). At the first step, $NiCo_2O_4$ microspheres were synthesized using a solvothermal method by taking $Ni(NO_3)_2 \cdot 6H_2O$ and $Co(NO_3)_2 \cdot 6H_2O$ as precursor solution with acetone and N,N-Dimethylformamide as solvents, and further phthalic acid had been added. This whole solution was transferred into a Teflon based autoclave and kept at 160°C for

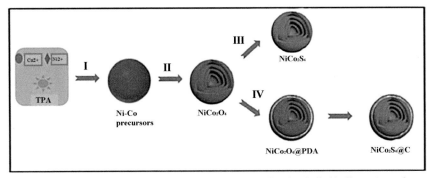

Figure 2.9: The fundamental synthetic routes of as-prepared multi-shelled hollow microspheres (reproduced with permission from Zuo et al. 2020).

24 hours. These $NiCo_2O_4$ microspheres further were dispersed into tris buffer solution with the addition of dopamine hydrochloride. The obtained composite was transferred to a quartz tube furnace, calcined at 500°C for four hours under H_2S atmosphere at a heating rate of 2°C min^{-1} to get multi-shelled $NiCo_2S_4$@C hollow microspheres. Figure 2.9 represents the schematic diagram of the synthesis of $NiCo_2S_4$@C hollow microspheres (Zuo et al. 2020). Our research group had also synthesized hierarchical and solid NiV_2S_4 nanospheres using template free process (Srivastava et al. 2020). The schematic diagram is shown in Figure 2.10 in which ethylene glycol and DMF are used as solvents, whereas $VOSO_4,5H_2O$ and $Ni(CH_3CO_2)_2.4H_2O$ taken as precursor solution. The solvothermal process was maintained at 170°C for five hours. In this process, the control factor was the concentration of thiourea and refluxing time. Figure 2.11 is showing the prominent transformation in the structure with a distinct change in the particle size between solid and hierarchical NPs.

Similar research work has been reported by Wang et al. in the preparation of Mn_3O_4/MnS heterostructures building multi-shelled hollow microspheres (Wang et al. 2020). In this work, the Mn-based coordination polymer spheres were synthesized using the solvothermal process of Mn^{2+} with an organic compound. The multi-shelled Mn_2O_3 spheres were synthesized through a pyrolysis process of this metal-based coordination polymer in the air atmosphere. Further, the Mn_2O_3 spheres went through the reaction with sulfur powder to form the final Mn_3O_4/MnS multi-shelled hollow spheres in which composition had been regulated by varying the content of sulfur powder during the sulfidation process (Wang et al. 2020). Ren et al. reported one sacrificial templating method to synthesize multi-shelled TiO_2 hollow spheres, which used carbonaceous microspheres as templates (Ren et al. 2014). Wu et al. synthesized a very easy method to produce multi-layered NiO hollow spheres (Wu et al. 2016). In this process,

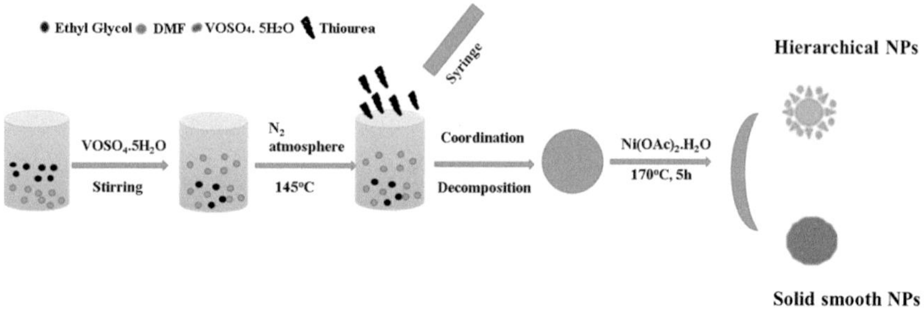

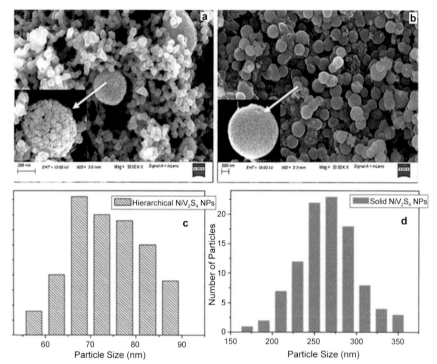

Figure 2.10: Schematic for one-pot synthesis of ternary hierarchical porous and solid NiV$_2$S$_4$ NPs (reproduced with permission from Srivastava et al. 2020).

Figure 2.11: SEM images of ternary (a) hierarchical NiV$_2$S$_4$ NPs, (b) solid NiV$_2$S$_4$ NPs, particle size histograms for (c) hierarchical NiV$_2$S$_4$ NPs and (d) solid NiV$_2$S$_4$ NPs (reproduced with permission from Srivastava et al. 2020).

they had reported a dehydration reaction between the surface functional groups (–OH and –C=O) on carbonaceous particles with metal oxide hydrate in solution, resulted in the formation of carbonaceous particle@ metal oxide hydrate core-shell structure. Similar work had been reported by many research groups to synthesize multi-shelled Mn$_2$O$_3$, α-Fe$_2$O$_3$,

Co_3O_4, ZnO and SnO_2 hollow spheres using carbonaceous core materials as the template (Wang et al. 2015, Xu et al. 2014, Wang et al. 2014, Dong et al. 2012, Qian et al. 2009). In this typical synthesis process, styrene droplets are considered as the soft template, whereas methyltriethoxysilane is taken as a hard one. Further, this silica shell is utilized as a nanoreactor. Some *in situ* PS chains present in the nanoreactor migrated to the outer surface of silica shells through strong capillary force. Simultaneously, part of siloxane oligomers migrated to the surface due to interfacial activity, resulted in the formation of hierarchical SOS. With consecutive calcination process of SOS produced the hollow sphere-on-sphere desired products (Wang et al. 2020). In the year 2018, Jang et al. reported one synthetic method using zeolitic imidazolate frameworks (ZIFs) on the surface of silica spheres by the Sonogashira coupling of tetra(4-ethynylphenyl) methane38 with 1,4-diiodobenzene and the successive etching of the inner silica (Jang et al. 2019). Metal-organic framework (MOF) was recently used to prepare advanced transition metal oxide electrodes due to the capability of forming a well-organized nanostructure. By thermal annealing MOF materials, carbon-coated transition metal oxide with the desired structure can be achieved in a facile manner (Jang et al. 2014). In most of the studies, the MOFs were formed by the solvothermal reactions with metal precursor solution, trimesic acid as organic ligand and polyvinyl pyrrolidone (PVP) (Zou et al. 2016). Figure 2.12 presents the schematic diagram of the synthesis process of $M_xO_y/M/$graphene multi-

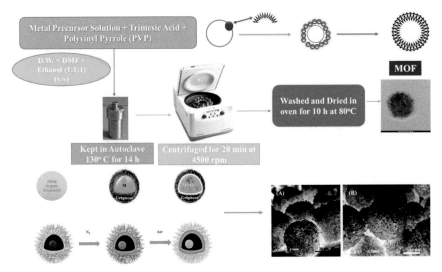

Figure 2.12: Schematic diagram of $M_xO_y/M/$graphene multi-layered hollow spherical nanomaterials (reproduced with permission from Kumar et al. 2020).

layered nano-spheres. The concentration of PVP controls the size, shape, porosity and crystallinity of the end products.

A modified arc-discharge method with an air-annealing process has been proved as another promising synthesis method to synthesize metal nano-particle encapsulated onion-like carbon nano-capsules (Chattopadhyay et al. 2020, Liu et al. 2015, Langea et al. 2003).

Our research group has recently reported on the arc-discharge method to form Mo-Ni encapsulated onion-like carbon nano-capsules presented in Figure 2.13 (Chattopadhyay et al. 2020). In this typical process, metal powders were ball-milled with graphite powder, which was further compacted into a cylinder-shaped structure under pressure and placed into one pit of a water-cooled carbon crucible. Here, carbon needle was considered as a cathode, whereas metal-mixtures employed as an anode target. In this modified method, liquid ethanol was introduced to the chamber.

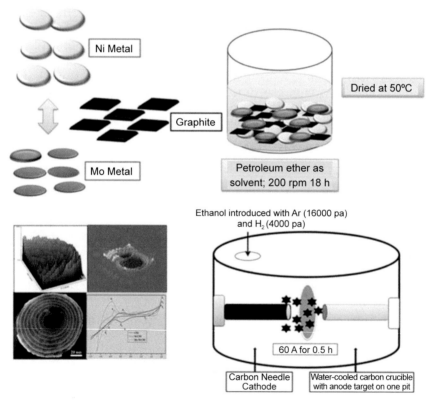

Figure 2.13: Schematic diagram of Mo-Ni encapsulated onion-like nano-capsules (reproduced with permission from Chattopadhyay et al. 2020).

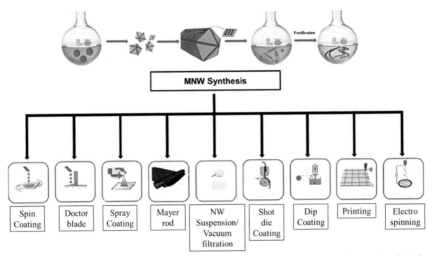

Figure 2.14: Schematic diagram of metal nano-wires synthetic strategies (reproduced with permission from da Silva et al. 2016).

5. Metal Oxide Nano-wires Synthetic Strategies

Metal nano-wires have been extensively attracted attention due to their simple solution-based synthesis of the metal nanowires with a very large surface area.

A wide range of methods was exploited to form MNWs, among which the most widely applied methodology in MNW synthesis is a solution-phase chemical synthesis reaction. As MNWs can be grown in a liquid phase at a lower range of temperature, the synthesis of MNWs may be scalable toward commercialization. In the MNW development with a traditional chemical synthesis route, the metal ion is initially reduced into metal nuclei, and further with the nuclei growing produces a nanosized metallic material with capping agents anchored on the specific material surface, which finally leads to the anisotropic assembly of nuclei into MNW rather than nanoparticles. During MNW formation the by-products and reaction agents, such as nanoparticles and capping agents, would generate an additional optical and electrical loss of MNWN electrode. Thus, it is critical to remove them to get quality MNWNs. So far, a variety of MNWs with potential performing characteristics for photovoltaic applications was reported, such as silver nanowires (Ag NWs) (da Silva et al. 2016, Wu et al. 2013, Coskun et al. 2011, Li et al. 2015, Zhang et al. 2017, Sim et al. 2016), copper nanowires (CuNWs) (Guo et al. 2013, Xu et al. 2015, Jin et al. 2011, Rathmell et al. 2011, Wang et al. 2018, Hwang et al. 2016, Ravi Kumar et al. 2015), gold nanowires (Au NWs) or metal alloy nanowires(core@shell NWs) (Niu et al. 2017, Rathmell et al. 2012, Song et al. 2014), etc. Among these MNWs, the most potentially applied as an electrode for the new-generation solar cells is Ag NWs and

Cu NWs due to their high conductivity, transparency, scalability and relatively low-cost (Song et al. 2014).

(a) Silver Nano-wires

Silver nanowires are one of the most widely studied MNWs as silver executes high electrical conductivity at room temperature and the formation of Ag NWs is facile, scalable and reproducible in nature. To date, the commercial silver NWs produce comparable or even better optical transparency and conductivity than that of the ITO electrode. Ag nanowires can be synthesized using different methodologies, such as the polyol method (da Silva et al. 2016, Wu et al. 2013, Coskun et al. 2011, Li et al. 2015, Zhang et al. 2017, Sim et al. 2016) and electrospinning (Wu et al. 2013). Among them, the most applied method for the AgNW formation is a polyol method (Sun et al. 2002). During the AgNW synthesis with such a method, ethylene glycol (EG) and polyvinylpyrrolidone (PVP) are used to transform Ag precursor (usually $AgNO_3$) into Ag NWs. In this case, PVP is widely utilized as a polymeric capping agent that results in synthesize anisotropic surface passivation on the NWs (Wiley et al. 2005, Tao et al. 2003). In brief, the ethylene glycol is preheated to the specific temperature ($\sim 150°C$) to produce glycolaldehyde and reduce Ag^+ into Ag^o and further, the Ag atom is agglomerated into the nucleus. The nucleus grows as a seed, which is further elongated to forma structure like a rod, i.e., nanorod. During the Ag nanorod growth, PVP preferentially adheres along the (100) surface, whereas the (111) surface of the Ag nanorod is free to grow. Therefore, the seed has no control in anisotropy growing into NWs along the direction of the (111) surface, and the NW surface is formed by five single-crystal tetrahedral subunits via sharing (111) twin planes and boundaries, producing a pentagonal pyramid shape (Sun et al. 2002). A typical high-pressure polyol method can form the Ag NWs with an aspect ratio of 50–300 with a diameter range of 20 nm and the length is up to ~ 25 mm (Lee et al. 2012). Other attempts have been devoted to forming ultra-long (> 50 mm) and ultra-thin (< 25 nm) Ag NWs with a high aspect ratio (> 2,000) and with by-products as few as possible (Li et al. 2015). One research group has reported a high-efficiency synthetic strategy by controlling the concentration of bromide ion in the reaction system to grow thinner Ag NWs and thereafter using a selective precipitation process to purify the Ag NWs (Li et al. 2015). In this case, AgNW exhibited a thin diameter of 20 ± 2 nm with a greater aspect ratio to the value of 2,000. Another research group further modified this process by a one-pot polyol approach via applying a syringe pump to slow introduction of $AgNO_3$ solution into the reaction agents, which diminished the diameters of the Ag NWs to below 20 nm (da Silva et al. 2016). In another work, the

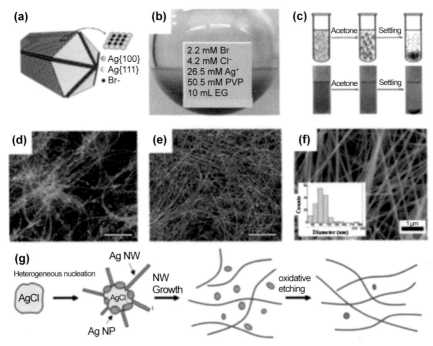

Figure 2.15: (a) Schematic illustrating the formation of Ag NWs. The PVP and Br⁻ can limit the lateral growth of Ag NW by capping the (100) facets on the side surface. Reproduced with permission (da Silva et al. 2016). Copyright 2016, American Chemical Society. (b) Photograph of reaction flask after the growth of Ag NWs. (c) Schematic illustrating the process for Ag NW purification. SEM images of Ag NWs (d) before and (e) after purification. Reproduced with permission (Li et al. 2015). Copyright 2015, American Chemical Society. (f) SEM images of Ag NW adjusted by Cl⁻ and Fe³⁺ dual controlling agents. Inset shows the statistical distribution of NWs diameter. Reproduced with permission (Zhan et al. 2016). Copyright 2016, The Royal Society of Chemistry. (g) Nucleation and growth mechanism of Ag NWs. Reproduced with permission (Sim et al. 2016). Copyright 2016, Wiley-VCH.

length Ag NWs increased up to ~ 220 mm with an aspect ratio of 4,000 via a one-pot hydrothermal method using PVP and FeCl₃ as the capping agents and then purified the NWs with a filter cloth (pore size was about 30X 50 μm²) (Zhang et al. 2017). Another research group demonstrated a one-pot stirring-free method to synthesize AgNWs with a tuneable length and diameter (D = 45–220 nm, L = 10–230 mm) by employing Cl⁻ and Fe³⁺ as dual controlling agents to adjust the reduction speed of Ag⁺ and restrict the nucleus formation (Zhan et al. 2016). Although, conventional long-chain PVP capping agent covers MNWs forming an insulating layer, and therefore drastically reduced the conductivity of the corresponding MNWN electrodes (Sim et al. 2016).

(b) Copper Nanowires

CuNWs is another potential alternative material for ITO electrode as it has comparable conductivity with Ag NWs and much lower cost than that of AgNWs (Rathmell et al. 2011, Jeong et al. 2008). The CuNWs can be developed by following different processes, viz., chemical vapor deposition (CVD) (Choi et al. 2004), template-assisted method (Gerein et al. 2005), electrospinning (Wu et al. 2010, Yang et al. 2017) and chemical synthesis (Rathmell et al. 2010, Cui et al. 2015). Among these methodologies, the solution phase chemical synthesis method has been considered as the simplest one, which has been successfully utilized for scalable production. Two general solution-phase approaches are applied to form copper nanowires including (i) ethylenediamine (EDA)-mediated process and (ii) alkylamine-mediated process. One research group had initiated the solution-phase approach of the CuNW synthesis in which diameter and length had been analyzedas 90–120 nm and 40–50 mm, respectively (Chang et al. 2005). In this typical process, the reaction proceeds under a strong basic condition (pH = 14–15) in which the Cu^{2+} ion is reduced by hydrazine (N_2H_4) and thereafter anisotropic grows along with one direction of Cu material (Kim et al. 2017). In this method, the role of EDA during the synthesis of CuNWs functions as a capping agent that actually promotes the anisotropic growth of NWs. In recent work, Kim et al. explained EDA's role as a "facet-selective promoter" during the NW growth instead of a capping agent (Kim et al. 2017). Another strategy to develop CuNW is an alkylamine-mediated reaction, which generally proceeds in a neutral or moderately basic medium at a higher temperature (175°C) (Ye et al. 2014).

Other metal nanowires have also been claimed as a potential candidate in various applications and their synthetic strategies have also become well accepted among the scientific community. Among them, AuNW has drawn huge attention in their successful synthesis, as they have been accepted as the most stable metals among others. To apply them in a specific application, the development of ultrathin AuNW becomes essential with a high aspect ratio value. In this typical method, chloroauric acid ($HAuCl_4$) and OAm are usually utilized as the basic reactants to synthesize Au NW with a diameter of less than 10 nm (Huo et al. 2008). Similar to the synthesis of CuNW, the role of OAm has been proved as both a capping and reducing agent in this case. Yang et al. formed ultrathin single-crystal AuNWs with a diameter of ~ 1.6 nm via aging the OAm solution of $HAuCl_4$ at room temperature (20–25°C) for four days long (Huo et al. 2008). They have demonstrated that with an increase in the aging time of the solution, the other shaped nano-products (nanorods or NPs) would be formed in a lesser amount (Halder et al. 2007). Additionally, the usage of other organic solvents [toluene (Halder et al. 2007) and hexane (Yu et al. 2016)] and with the application of reduction agents [ascorbic acid (Halder et al. 2007) and silane (Yu et al. 2016)] can effectively reduce the overall reaction time. In a work, a large amount of ultrathin Au NWs have been obtained with a diameter of 2 n min 12 hours with the addition of tris(trimethylsilyl)silane into the hexane solution of $HAuCl_4$ and OAm. To enhance the conductivity and optical transparency of the AuNWs, one research group synthesized AuNWs with a length of ~ 500 nm and an aspect ratio larger than 1,000 by applying an *in situ* pattern reaction, and the high aspect ratio leads to a higher conductivity (15 U sq^{-1}) and transmittance (~ 78%) (Bao et al. 2016). Chen et al. developed vertically aligned ultrathin (d ¼ 6 nm) AuNWs under ambient conditions with the application of a thiolated ligand capping agent of 4-mercaptobenzoic acid (4-MBA) and L-ascorbic acid reduction agent (Wang et al. 2017). Moreover, Au can also be utilized as an inert shell of CuNW to prevent its oxidation. Although the high cost of Au restricts their application for this purpose. In spite of the advantages of Ag and Cu NWs, poor electrical stability controls their independent application as the principal material for high-performance energy devices. In avoidance of such a problem, scientists synthesized the NWs by using alloys or core-shell structures which generates the synergistic effects between these components. Typically, the core-shell NWs are generally formed by the incorporation of pristine MNWs into an aqueous solution that contains metal ion precursors, capping agents, reducing agents or other auxiliary agents. Yeo et al. developed electrochemically stable Ag-Au core-shell NWs with high durability and transparency (> 85%) through the liquid phase reduction of Au+ precursor onto the surface of the AgNWs, and the thickness of Au coating can be tuned by adjusting pH and precursor injection speed (Moon et al. 2017).

6. Metal Hollow Spheres as Gas Sensors

Chemo-resistive n-type oxide semiconductors, such as SnO_2, ZnO, TiO_2, In_2O_3 and WO_3, have been widely used to detect explosive, toxic and harmful gases (Kaur et al. 2020). Hollow structures are promising nano-architectures for the applications of gas sensors on account of their high surface area and gas accessible configurations of thin shells (Lee et al. 2009, Kim et al. 2011). Not only the outer surfaces but also the inner ones participate in the gas sensing reaction. In general, oxide hollow structures are prepared by applying a coating of metal precursors onto polymeric spheres and subsequent removal of sacrificial templates by heat treatment (Lou et al. 2007, Caruso et al. 2001).

In 2011, Kim et al. reported In_2O_3 hollow spheres as gas sensing materials. Which were synthesized using a hydrothermal methodology (Kim et al. 2011). The average diameters of ~ 100 In_2O_3, $Sb-In_2O_3$, $Cu-In_2O_3$, $Nb-In_2O_3$, $Pd-In_2O_3$ and $Ni-In_2O_3$ spheres were 2.3 ± 0.5 µm, 2.4 ± 0.7 µm, 2.2 ± 0.4 µm, 2.3 ± 0.6 µm, 2.3 ± 0.5 µm and 2.2 ± 0.5 µm, respectively. The decrease of sphere diameters during heat treatment can be attributed to the shrinkage of spheres by the decomposition of carbon cores.

At the sensor temperature of 371°C, the gas responses of the In_2O_3 sensor to 100 ppm NH_3, 5 ppm H_2S, 100 ppm H_2, 100 ppm CO and 500 ppm CH_4 ranged from 2.0 to 4.6. All the gas responses were decreased by the loading of Nb and Sb. In contrast, the loading of Cu increased the responses to all the gases. In the $Cu-In_2O_3$ sensor, although the response to H2S (7.5) was higher than the response to the other four gases, it was not markedly higher than the response to NH_3 (6.2). Among the five different sensors, the loading of Pd showed the most selective detection of H_2S. The H_2S response of the $Pd-In_2O_3$ sensor was 8.7, while the responses to NH_3, H_2, CO, and CH_4 were 5.1, 3.0, 3.9 and 3.2, respectively. The selectivity to a specific gas was defined as "SSG/SIG" (SSG: gas response to a specific gas, SIG: gas response to interference gas). The $SH2S/SIG$ values ranged from 1.7 to 5.7. The loading of Ni increased all the gas responses. In particular, the response to CO was enhanced to a great extent. The SCO/SIG values ranged from 1.6 to 2.2. Thus, the high response to CO (12.9) with the lower cross-responses to NH_3, H_2S, H_2 and CH_4 (5.8–8.3) demonstrates that the Ni-In_2O_3 sensor can be used for selective CO detection. The selectivity of the gas sensing reaction was also influenced by the variation of sensor temperature. When the sensor temperature was increased to 440°C, the gas responses of all the sensors tended to increase. In the pure In_2O_3 sensor, the NH_3 response (10.6) was the highest whereas the response to H_2S (7.2) was comparable. The loadings of Sb and Nb led to a decrease in gas responses. Similar work with WO_3 hollow spheres was reported by another group which had been utilized as sensors with good sensitivity to alcohol, acetone, CS_2 and other organic gases (Chakraborty et al. 2006).

Another research group developed ZnO hollow spheres with their potential sensitivity characteristics in NH_3 and NO_2 detection (Zhang et al. 2009). The $\alpha-Fe_2O_3$ hollow balls had also been reported with their remarkable activity in sensing NH_3 and CO at the temperature range of 250 to 450°C (Huang et al. 2016). In 2017, Ku et al. had reported on ZnO/$ZnCo_2O_4$ hollow spheres, which were shown high selectivity toward acetone gas with notable sensitivity (S = 69% to 5 ppm at 250°C) (Koo et al. 2017). These spheres were synthesized through MOF driven methodology. Recently, Zhao et al. developed WO_3 hollow spherical materials, which had exhibited good sensing properties to sub-ppm NO_2 at RT, namely a high response, low detection limit, good selectivity and repeatability (Zhao et al. 2020).

7. Metal Nanowires as Gas Sensors

For n-type NWs that are widely used in sensing applications, variations of the conduction channel due to the presence of the target gas are responsible for sensing signals. Initially, in the air, an electron-depletion layer is formed on the outer surfaces of the NW due to the extraction of electrons by adsorbed oxygen ions. This causes the formation of a so-called conduction channel inside the core of NW. Then, depending on the interaction with either oxidizing or reducing gases, the conduction channel either contracts or expands, respectively, leading to either an increase or a decrease of NW resistance and the appearance of a sensing signal. Depending on the ratio between the characteristic transverse sizes of the nanostructures, there are two main electron-transport regimes (i) due to the presence of the adsorbed oxygen species, the number of free electrons is reduced at the oxide surface states and the electron transport is modulated by potential barriers. Thus, the gas-sensing mechanism is dependent on both the width and height of the contact potential barriers. (ii) The Fermi level is controlled by surface states under flat energy bands, and the mechanism depends mainly on the release of electrons from the surface states to the conduction band. Therefore, the influence of potential barriers, which are most significant at the contacts between agglomerates, can be neglected (Kuprianov et al. 1996, Williams et al. 2000).

The sensing properties of single NWs are affected by the NW diameter, the NW synthesis procedure and there actions that occur on the NW surfaces. The realization of single-NW gas sensors has significant fabrication issues, e.g., the formation of electrical contacts. Electron-beam lithography can be used to define the electrical contacts for a single-NW gas sensor. However, a simpler method is synthesis, sonification and dispersal of the NW on another substrate equipped with electrodes. Because of the complex fabrication processes, the commercialization of single-NW gas sensors is very difficult (Choi et al. 2008). Tonezzer and Hieu (Tonezzer et al. 2012) prepared monocrystalline SnO_2 NWsvia CVD. They were subsequently dispersed onto a substrate, and by applying electrical contacts, single-NW gas sensors with different diameters were fabricated for NO2-sensing studies. The foregoing results were confirmed by Lupan et al. (2010). The responses of single ZnO NW (100 and 200 nm in diameter) gas sensors were directly related to the diameter of the NW, and the highest gas response was observed for a sensor with an NW diameter of 100 nm. In a gas sensor with a smaller NW diameter, the diameter and Debye length were comparable, resulting in significant resistance modulation and a higher response (Lupan et al. 2010). In another study, the responses of single SnO2 NW gas sensors with different diameters (20–140 nm) were proportional to the inverse of the diameter; the smallest NW diameter yielded the highest gas response (Asadzadeh et al. 2019). In general, single-NW gas sensors have inferior sensing properties to multiple- and networked-NW gas sensors. SnO_2 NWs were reported by Bang et al. (2018). The SnO_2 NWs and Bi_2O_3 branches were simultaneously produced via the VLS method. The sensor exhibited a high response of 56.92 to 2-ppm NO_2 gas. The high sensing performance of the branched-NW sensor was related to the large surface area of the sensor resulting from the Bi_2O_3 branching and the formation of Bi_2O_3-Bi_2O_3 and SnO_2-SnO_2 homojunctions and Bi_2O_3-SnO_2 heterojunctions. CuO-functionalized SnO_2 branched NWs were realized via a single process by using Cu as a catalyst for the growth of SnO_2 branches and the source of CuO NWs. Using this method, Kim et al. (2012) fabricated NWs CuO-functionalized SnO_2 branched NWs. After annealing at 700°C, CuO was located at the tips of the SnO_2 branches. When the gas sensor was exposed to H_2S, a high response was observed, which was mainly related to the conversion of semiconductive CuO into metallic CuS, which induced significant resistance modulation. However, because the CuO-induced depletion region covered a considerable volume of the SnO_2 branches, there were no electrons available for incoming NO_2 gas molecules, resulting in a low response to NO_2 gas (Kim et al. 2012). Kim et al. prepared ZnO branched SnO_2 NWs via a VLS growth method and then sputtered NWs with a Co shell, followed by annealing treatment to convert the Co layer into Co NPs at the surfaces of the ZnO branches (Kim et al. 2015).

8. Metal Nanotubes as Gas Sensors

In recent times, the most popular gas sensors are usually consisting of surface-controlled resistance sensors in which metal oxide semiconductors mainly been considered semiconductor resistance gas sensors (Jeong et al. 2019). The material's surface shows strong adsorption characteristics and high reactivity due to the presence of free electrons in huge numbers under a conduction band, and it may also attribute to the appearance of oxygen vacancies on the surface of the metal semiconductors, which transforms under the action of surface gas. Thus, the electrical parameters have been considered for measurement purposes (Rabee et al. 2019). The main reason for the popularity of metal oxides is their outstanding physical and chemical properties. Additionally, they are very economical to synthesize with very simple methods to produce (Wang et al. 2019).

Song et al. reported the development of hollow porous core-shell NiO nanotubes, which were synthesized using a hydrothermal method (Song et al. 2011). These materials possessed an enormous number of micropores with a high specific surface of 97.3 m^2.g^{-1}. The high specific surface area resulted in the greater sensitivity of the NiO gas sensor to 50 ppm toward ethanol gas. The hollow core-shell structure of these materials enhanced the diffusion quality for ethanol molecules, which resulted in the rapid transport into the interior of the sensor, enabled the materials into potential gas sensors. Ma et al. developed hollow microtubules of In$_2$O$_3$ with degreased cotton as a soft biological template (Ma et al. 2019). These In$_2$O$_3$ hollow micro-tubes were formed with a length of 50–70 μm, the width was 5–6 μm and the wall thickness was about 1 μm. The R$_{gas}$/R$_{air}$ value to 10 ppm Cl$_2$ gas was 1,051 for the hollow microtubule sensor of In$_2$O$_3$, which was 25 times higher than that of In$_2$O$_3$ particles at 200°C. This remarkable performance was affirmed by the hollow morphology and high specific area of these nanotubes, which were consisting of excellent oxygen vacancy and narrow band gap. In another work, Wanit et al. developed the ZnO nano-trees with a multi-stage branching structure through a hydrothermal growth method in which numerous nano-trees with dense distribution formed a vast nano forest (Wanit et al. 2012). The nano forest structured materials were performed remarkably as dye-sensitized cells. They worked in the same manner as a forest in which solar energy converted into electricity in successful manner. Obviously, this kind of nano-forest structure had a very large specific surface area, which could be used to significantly improve the gas-sensing performance of materials. Moreover, these experiments had shown that the light conversion efficiency had improved manyfold and further the overall light conversion efficiency improved five times higher in comparison to those of ZnO nanowires.

References

Asadzadeh, M.Z., A. Köck, M. Popov, S. Steinhauer, J. Spitaler, L. Romaner. 2019. Response modeling of single SnO$_2$ nanowire gas sensors. *Sens. Actuators B Chem.* 295, 22–29.

Aswal, D.K., A. Singh, S. Sen, M. Kaur, C.S. Viswandham, G.L. Goswami, S.K. Gupta. 2002. Effect of grain boundaries on paraconductivity of YBa$_2$Cu$_3$O$_x$. *J. Phys. Chem. Solids* 63, 1797–1803.

Aswal, D.K., N. Joshi, A.K. Debnath, K.P. Muthe, S.K. Gupta, J.V. Yakhmi, D. Vuillaume. 2005. Morphology-dependent electric transport in textured ultrathin Al films grown on Si. *J. Appl. Phys.* 98, doi: 10.1063/1.1977188.

Bang, J.H., M.S. Choi, A. Mirzaei, Y.J. Kwon, S.S. Kim, T.W. Kim, H.W. Kim. 2018. Selective NO$_2$ sensor based on Bi$_2$O$_3$ branched SnO$_2$ nanowires. *Sens. Actuators B Chem.* 274, 356–369.

Bao, C., W. Zhu, J. Yang, F. Li, S. Gu, Y. Wang, T. Yu, J. Zhu, Y. Zhou, Z. Zou. 2016. Highly flexible self-powered organolead trihalide perovskite photodetectors with gold nanowire networks as transparent electrodes. *ACS Appl. Mater. Interfaces* 8, 23868–23875.

Barsan, N., U. Weimar. 2001. Conduction model of metal oxide gas sensors. *J. Electroceram.* 7, 143–167.

Bhandarkar, V., S. Sen, K.P. Muthe, M. Kaur, M.S. Kumar, S.K. Deshpande, S.K. Gupta, J.V. Yakhmi, V.C. Sahni. 2006. Effect of deposition conditions on the microstructure and gas-sensing characteristics of Te thin films. *Mat. Sci. Eng. B* 131, 156–161.

Caruso, F., R.A. Caruso, H. Möhwald. 1998. Nanoengineering of inorganic and hybrid hollow spheres by colloidal templating. *Science* 282, 1111–1114.

Caruso, F., X. Shi, R.A. Caruso, A. Susha. 2001. Hollow titania spheres from layered precursor deposition on sacrificial colloidal core particles. *Adv. Mater.* 13, 740–744.

Chakraborty, S., A. Sen, H.S. Maiti. 2006. Selective detection of methane and butane by temperature modulation in iron doped tin oxide sensors. *Sens. Actuat. B* 115, 610–613.

Chang, Y., M.L. Lye, H.C. Zeng. 2005. Large-scale synthesis of high-quality ultralong copper nanowires. *Langmuir* 21, 3746–3748.

Chattopadhyay, J., S. Singh, R. Srivastava. 2020. Enhanced electrocatalytic performance of Mo–Ni encapsulated in onion-like carbon nano-capsules. *J. Appl. Electrochem.* 50, 207–216.

Choi, H., S.H. Park. 2004. Seedless growth of free-standing copper nanowires by chemical vapor deposition. *J. Am. Chem. Soc.* 126, 6248.

Choi, Y.J., I.-S. Hwang, J.G. Park, K.J. Choi, J.H. Park, J.-H. Lee. 2008. Novel fabrication of an SnO_2 nanowire gas sensor with high sensitivity. *Nanotechnology* 19, 095508.

Cole, M.T., R.J. Parmee, W.I. Milne. 2016. Nanomaterial-based x-ray sources. *Nanotechnology* 27, 082501.

Coskun, S., B. Aksoy, H.E. Unalan. 2011. Polyol synthesis of silver nanowires: an extensive parametric study. *Cryst. Growth Des.* 11, 4963–4969.

Cui, F., Y. Yu, L.T. Dou, J.W. Sun, Q. Yang, C. Schildknecht, K. Schierle-Arndt, P.D. Yang. 2015. Synthesis of ultrathin copper nanowires using tris(trimethylsilyl)silane for high-performance and low-haze transparent conductors. *Nano Lett.* 15, 7610–7615.

Cui, Z., Y. Han, Q. Huang, J. Dong, Y. Zhu. 2018. Electrohydrodynamic printing of silver nanowires for flexible and stretchable electronics. *Nanoscale* 10, 6806–6811.

da Silva, R.R., M. Yang, S.I. Choi, M. Chi, M. Luo, C. Zhang, Z.Y. Li, P.H. Camargo, S.J. Ribeiro, Y. Xia. 2016. Facile synthesis of sub-20 nm silver nanowires through a bromide-mediated polyol method. *ACS Nano.* 10, 7892–7900.

Dong, Z., X. Lai, J.E. Halpert, N. Yang, L. Yi, J. Zhai, D. Wang, Z. Tang, L. Jiang. 2012. Accurate control of multishelled ZnO hollow microspheres for dye-sensitized solar cells with high efficiency. *Adv. Mater.* 24, 1046–1049.

Eslamian, M. 2017. Excitation by acoustic vibration as an effective tool for improving the characteristics of the solution-processed coatings and thin films. *Prog. Org. Coating* 113, 60–73.

Gerein, N.J., J.A. Haber. 2005. Effect of ac electrodeposition conditions on the growth of high aspect ratio copper nanowires in porous aluminum oxide templates. *J. Phys. Chem. B* 109, 17372–17385.

Gu, X., C. Li, X. Liu, J. Ren, Y. Wang, Y. Guo, Y. Guo, G. Lu. 2009. Synthesis of nanosized multilayered silica vesicles with high hydrothermal stability. *J. Phys. Chem. C* 113, 6472–6479.

Guo, H., N. Lin, Y. Chen, Z. Wang, Q. Xie, T. Zheng, N. Gao, S. Li, J. Kang, D. Cai, D.L. Peng. 2013. Copper nanowires as fully transparent conductive electrodes. *Sci. Rep.* 3, 2323–2326.

Halder, A., N. Ravishankar. 2007. Ultrafine single-crystalline gold nanowire arrays by oriented attachment. *Adv. Mater.* 19, 1854–1858.

Henrich, V.E., P.A. Cox. 1994. The Surface Science of Metal Oxides; Cambridge University Press: Cambridge, UK.

Huang, C.C., W. Huang, C.S. Yeh. 2011. Shell-by-shell synthesis of multi-shelled mesoporous silica nanospheres for optical imaging and drug delivery. *Biomaterials* 32, 556–564.

Huang, X.J., J.H. Liu, D.L. Shao, Z.X. Pi, Z.L. Yu. 2003a. Rectangular mode of operation for detecting pesticide residue by using a single SnO_2-based gas sensor. *Sens. Actuat. B* 96, 630–635.

Huang, X.J., J.H. Liu, Z.X. Pi, Z.L. Yu. 2003b. Detecting pesticide residue by using modulating temperature over a single SnO_2-based gas sensor. *Sensors* 3, 361–370.

Huang, X.J., F.L. Meng, Z.X. Pi, W.H. Xu, J.H. Liu. 2004a. Gas sensing behavior of a single tin dioxide sensor under dynamic temperature modulation. *Sens. Actuat. B* 99, 444–450.

Huang, X.J., Y.F. Sun, F.L. Meng, J.H. Liu. 2004b. New approach for the detection of organophosphorus pesticide in cabbage using $SPME/SnO_2$ gas sensor: Principle and preliminary experiment. *Sens. Actuat. B* 102, 235–240.

Huang, X.J., L.C. Wang, Y.F. Sun, F.L. Meng, J.H. Liu. 2004c. Quantitative analysis of pesticide residue based on the dynamic response of a single SnO_2 gas sensor. *Sens. Actuat. B* 99, 330–335.

Huang, X.J., F.L. Meng, Y.F. Sun, J.H. Liu. 2005. Study of factors influencing dynamic measurements based on SnO_2 gas sensor. *Sensors* 4, 95–104.

Hung, C.M., N.D. Hua, N.V. Duy, N.V. Toan, D.T.T. Le, N.V. Hieu. 2016. Synthesis and gas-sensing characteristics of α-Fe_2O_3 hollow balls. *J. Sci. Adv. Mater. Dev.* 1, 45–50.

Huo, Z., C.K. Tsung, W. Huang, X. Zhang, P. Yang. 2008. Sub-two nanometer single crystal Au nanowires. *Nano Lett.* 8, 2041–2044.

Hwang, C., J. An, B.D. Choi, K. Kim, S.W. Jung, K.J. Baeg, M.G. Kim, K.M. Ok, J. Hong. 2016. Controlled aqueous synthesis of ultra-long copper nanowires for stretchable transparent conducting electrode. *J. Mater. Chem. C* 4, 1441–1447.

Jang, J.Y., T.M.D. Le, J.H. Ko, Y.J. Ko, S.M. Lee, H.J. Kim, J.H. Jeon, T. Thambi, D.S. Lee, S.U. Son. 2019. Triple-, double-, and single-shelled hollow spheres of sulfonated microporous organic network as drug delivery materials. *Chem. Mater.* 31, 300–304.

Jeong, S., K. Woo, D. Kim, S. Lim, J.S. Kim, H. Shin, Y.N. Xia, J. Moon. 2008. Controlling the thickness of the surface oxide layer on Cu nanoparticles for the fabrication of conductive structures by ink-jet printing. *Adv. Funct. Mater.* 18, 679–686.

Jeong, H.I., S. Park, H.I. Yang, W. Choi. 2019. Electrical properties of $MoSe_2$ metal-oxide-semiconductor capacitors. *Mater. Lett.* 253, 209–212.

Jin, M., G. He, H. Zhang, J. Zeng, Z. Xie, Y. Xia. 2011. Shape-controlled synthesis of copper nanocrystals in an aqueous solution with glucose as a reducing agent and hexadecylamine as a capping agent. *Angew. Chem. Int. Ed.* 50, 10560–10566.

Kanazawa, E., G. Sakai, K. Shimanoe, Y. Kanmura, Y. Teraoka, N. Miura, N. Yamazoe. 2001. Metal oxide semiconductor N2O sensor for medical use. *Sens. Actuators B* 77, 72–77.

Kaur, M., S.K. Gupta, C.A. Betty, V. Saxena, V.R. Katti, S.C. Gadkari, J.V. Yakhmi. 2005. Detection of reducing gases by SnO2 thin films: An impedance spectroscopy study. *Sens. Actuat. B* 107, 360–365.

Kaur, N., M. Singh, E. Comini. 2020. One-dimensional nanostructured oxide chemoresistive sensors. *Langmuir* 36, 6326–6344.

Khare, N., D.P. Singh, A.K. Gupta, S. Sen, D.K. Aswal, S.K. Gupta, L.C. Gupta. 2005. Direct evidence of weak-link grain boundaries in a polycrystalline MgB_2 superconductor. *J. Appl. Phys.* 97, doi: 10.1063/1.1861506.

Kim, M.J., P.F. Flowers, I.E. Stewart, S. Ye, S. Baek, J.J. Kim, B.J. Wiley. 2017. Ethylenediamine promotes Cu nanowire growth by inhibiting oxidation of Cu(111). *J. Am. Chem. Soc.* 139, 277–284.

Kim, H.J., K.I. Choi, A. Pan, I.D. Kim, H.R. Kim, K.M. Kim, C.W. Na, G. Cao, J.H. Lee. 2011. Template-free solvothermal synthesis of hollow hematite spheres and their applications in gas sensors and Li-ion batteries. *J. Mater. Chem.* 21, 6549–6555.

Kim, H.W., H.G. Na, Y.J. Kwon, H.Y. Cho, C. Lee. 2015. Decoration of Co nanoparticles on ZnO-branched SnO_2 nanowires to enhance gas sensing. *Sens. Actuators B Chem.* 219, 22–29.

Kim, S.J., I.S. Hwang, Y.C. Kang, J.H. Lee. 2011. Design of selective gas sensors using additive-loaded In_2O_3 hollow spheres prepared by combinatorial hydrothermal reactions. *Sensors* 11, 10603–10614.

Kim, S.S., H.G. Na, S.W. Choi, D.S. Kwak, H.W. Kim. 2012. Novel growth of CuO-functionalized, branched SnO_2 nanowires and their application to H_2S sensors. *J. Phys. D Appl. Phys.* 45, 205301.

Kohl, D. 1989. Surface processes in the detection of reducing gases with SnO2-based devices. *Sens. Actuat.* 18, 71–113.

Koo, W.T., S.J. Choi, J.S. Jang, I.D. Kim. 2017. Metal-organic framework templated synthesis of ultrasmall catalyst loaded $ZnO/ZnCo_2O_4$ hollow spheres for enhanced gas sensing properties. *Sci. Rep.* 7, 45074–45079.

Kupriyanov, Y. 1996. Semiconductor Sensors in Physio-Chemical Studies. 1st ed.

Lai, X., J. Li, B.A. Korgel, Z. Dong, Z. Li, F. Su, J. Du, D. Wang. 2011. General synthesis and gas-sensing properties of multiple-shell metal oxide hollow microspheres. *Angew. Chem. Int. Ed.* 50, 2738–2741.

Langea, H., M. Siodaa, A. Huczkoa, Y.Q. Zhub, H.W. Krotob, D.R.M. Waltonb. 2003. Nanocarbon production by arc discharge in water. *Carbon* 41, 1617–1623.

Lee, J., P. Lee, H. Lee, D. Lee, S.S. Lee, S.H. Ko. 2012. Very long Ag nanowire synthesis and its application in a highly transparent, conductive and flexible metal electrode touch panel. *Nanoscale* 4, 6408–6414.

Lee, J.-H. 2009. Gas sensors using hierarchical and hollow oxide nanostructures: Overview. *Sens. Actuat. B* 140, 319–336.

Li, B., S. Ye, I.E. Stewart, S. Alvarez, B.J. Wiley. 2015. Synthesis and purification of silver nanowires to make conducting films with a transmittance of 99%. *Nano Lett.* 15, 6722–6726.

Li, Z., X. Lai, H. Wang, D. Mao, C. Xing, D. Wang. 2009. General synthesis of homogeneous hollow core-shell ferrite microspheres. *J. Phys. Chem. C* 113, 2792–2797.

Li, Z.P., Q.Q. Zhao, W.L. Fan, J.H. Zhan. 2011. Porous SnO_2 nanospheres as sensitive gas sensors for volatile organic compounds detection. *Nanoscale* 3, 1646–1652.

Liang, J., K. Tong, Q. Pei. 2006. A water-based silver-nanowire screen-print ink for the fabrication of stretchable conductors and wearable thin-film transistors. *Adv. Mater.* 28, 5986–5996.

Liu, X., S. Cheng, H. Liu, S. Hu, D. Zhang, H.A. Ning. 2012. Survey on gas sensing technology: Review. *Sensors* 12, 9635–9665.

Liu, X., C. Cui, N. Wu, S.W. Orb, N. Bi. 2015. Core/shell-structured nickel cobaltite/onion-like carbon nanocapsules as improved anode material for lithium-ion batteries. *Ceram. Int.* 41, 7511–7518.

Liu, J.H., X.J. Huang, F.L. Meng. 2007. The dynamic measurement of SnO2 gas sensor and their applications. pp. 177–214. *In*: Aswal, D.K., S.K. Gupta (eds.). Science and Technology of Chemiresistor Gas Sensors. Nova Science Publishers: New York, NY, USA.

Lou, X.W., C. Yuan, L.A. Archer. 2007. Shell-by-shell synthesis of tin oxide hollow colloids with nanoarchitectured walls: cavity size tuning and functionalization. *Small* 3, 261–265.

Lupan, O., V.V. Ursaki, G. Chai, L. Chow, G.A. Emelchenko, I.M. Tiginyanu, A.N. Gruzintsev, A.N. Redkin. 2010. Selective hydrogen gas nanosensor using individual ZnO nanowire with fast response at room temperature. *Sens. Actuators B Chem.* 144, 56–66.

Ma, J.W., H.Q. Fan, N. Zhao, W.M. Zhang, X.H. Ren, C. Wang. 2019. Synthesis of In_2O_3 hollow microspheres for chlorine gas sensing using yeast as bio-template. *Ceram. Int.* 45, 9225–9230.

Mao, D., J. Wan, J. Wang, D. Wang. 2018. Sequential templating approach: A ground breaking strategy to create hollow multishelled structures. *Adv. Mater.* 31, 1802874–1802892.

Meng, F.L., Y. Jia, J.Y. Liu, M.Q. Li, Y.F. Sun, J.H. Liu, X.J. Huang. 2010. Nanocomposites of sub-10 nm SnO_2 nanoparticles and MWCNTs for detection of aldrin and DDT. *Anal. Meth.* 2, 1710–1714.

Meng, X., F. Nawaz, F.S. Xiao. 2009. Templating route for synthesizing mesoporous zeolites with improved catalytic properties. *Nano Today* 4, 292–301.

Moon, H., J. Kwon, Y.D. Suh, D.K. Kim, I. Ha, J. Yeo, S. Hong, S.H. Ko. 2017. Ag/Au/polypyrrole core-shell nanowire network for transparent, stretchable and flexible supercapacitor in wearable energy devices. *Sci. Rep.* 7, 41981–41988.

Nazemi, H., A. Joseph, J. Park, A. Emadi. 2019. Advanced micro- and nano-gas sensor technology: A review. *Sensors* 19, 1285–1308.

Nguyen, V.D., V.H. Nguyen, T.H. Pham, D.C. Nguyen, M. Thamilselvan, J. Yi. 2008. Mixed SnO_2/TiO_2 included with carbon nanotubes for gas-sensing application. *Phys. E* 41, 258–263.

Niu, Z., F. Cui, Y. Yu, N. Becknell, Y. Sun, G. Khanarian, D. Kim, L. Dou, A. Dehestani, K. Schierle-Arndt, P. Yang. 2017. Ultrathin epitaxial Cu@Au core–shell nanowires for stable transparent conductors. *J. Am. Chem. Soc.* 139, 7348–7354.

Ogawa, H., M. Nishikawa, A. Abe. 1982. Hall measurement studies and an electrical-conduction model of tin oxide ultrafine particle films. *J. Appl. Phys.* 53, 4448–4455.

Park, J., J. Lee, Y. Noh, K.H. Shin, D. Lee. 2016. Flexible ultraviolet photodetectors with ZnO nanowire networks fabricated by large area controlled roll-to-roll processing. *J. Mater. Chem. C* 4, 7948–7958.

Qian, J.F., P. Liu, Y. Xiao, Y. Jiang, Y. Cao, X. Ai, H. Yang. 2009. TiO_2-coated multilayered SnO_2 hollow microspheres for dye-sensitized solar cells. *Adv. Mater.* 21, 3663–3667.

Rabee, A.S.H., M.F.O. Hameed, A.M. Heikal, S.S.A. Obayya. 2019. Highly sensitive photonic crystal fiber gas sensor. *Optik* 188, 78–86.

Rathmell, A.R., S.M. Bergin, Y.L. Hua, Z.Y. Li, B.J. Wiley. 2010. The growth mechanism of copper nanowires and their properties in flexible, transparent conducting films. *Adv. Mater.* 22, 3558–3563.

Rathmell, A.R., B.J. Wiley. 2011. The synthesis and coating of long, thin copper nanowires to make flexible, transparent conducting films on plastic substrates. *Adv. Mater.* 23, 4798–4803.

Rathmell, A.R., M. Nguyen, M. Chi, B.J. Wiley. 2012. Synthesis of oxidation-resistant cupronickel nanowires for transparent conducting nanowire networks. *Nano Lett.* 12, 3193–3199.

Ravi Kumar, D.V., K. Woo, J. Moon. 2015. Promising wet chemical strategies to synthesize Cu nanowires for emerging electronic applications. *Nanoscale* 7, 17195–17210.

Ren, H., R. Yu, J. Wang, Q. Jin, M. Yang, D. Mao, D. Kisailus, H. Zhao, D. Wang. 2014. Multishelled TiO_2 hollow microspheres as anodes with superior reversible capacity for lithium ion batteries. *Nano Lett.* 14, 6679–6684.

Rothschild, A., Y. Komem. 2004. The effect of grain size on the sensitivity of nanocrystalline metal-oxide gas sensors. *J. Appl. Phys.* 95, 6374–6380.

Sim, H., S. Bok, B. Kim, M. Kim, G.H. Lim, S.M. Cho, B. Lim. 2016. Organic-stabilizer-free polyol synthesis of silver nanowires for electrode applications. *Angew. Chem. Int. Ed.* 55, 11814–11818.

Soltani, R., A. Marjani, R. Soltani, S. Shirazian. 2020. Hierarchical multi-shell hollow micro–meso–macroporous silica for Cr(VI) adsorption. *Sci. Rep.* 10, 9788–9799.

Song, J., J. Li, J. Xu, H. Zeng. 2014. Superstable transparent conductive $Cu@Cu_4Ni$ nanowire elastomer composites against oxidation, bending, stretching, and twisting for flexible and stretchable optoelectronics. *Nano Lett.* 14, 6298–6305.

Song, X.F., L. Gao, S. Mathur. 2011. Synthesis, characterization, and gas sensing properties of porous nickel oxide nanotubes. *J. Phys. Chem. C* 115, 21730–21735.

Srivastava, R., J. Chattopadhyay, R. Patel, S. Agrawal, S. Nouseen, S. Kumar, S. Karmakar. 2020. Highly efficient ternary hierarchical NiV_2S_4 nanosphere as hydrogen evolving electrocatalyst. *Int. J. Hydrog. Energy* 45, 21308–21318.

Sun, Y.G., B. Gates, B. Mayers, Y.N. Xia. 2002. Crystalline silver nanowires by soft solution processing. *Nano Lett.* 2, 165–168.

Sun, Y.F., X.J. Huang, F.L. Meng, J.H. Liu. 2004. Study of influencing factors of dynamic measurements using SnO2 gas sensor. *Sens. Mater.* 17, 29–38.

Sun, Y.F., S.B. Liu, F.L. Meng, J.Y. Liu, Z. Jin, L.T. Kong, J.H. Liu. 2012. Metal oxide nanostructures and their gas sensing properties: A review. *Sensors* 12, 2610–2631.

Tao, A., F. Kim, C. Hess, J. Goldberger, R.R. He, Y.G. Sun, Y.N. Xia, P.D. Yang. 2003. Langmuir-blodgett silver nanowire monolayers for molecular sensing using surface-enhanced raman spectroscopy. *Nano Lett.* 3, 1229–1233.

Thomas, J.M. 2015. Sir Humphry Davy and the coal miners of the world: A commentary on Davy (1816) an account of an invention for giving light in explosive mixtures of fire-damp in coal mines. *Philos. Trans. A Math. Phys. Eng. Sci.* 373, 20140288–20140299. Doi: 10.1098/rsta.2014.0288.

Tomchenko, A.A., G.P. Harmer, B.T. Marquis, J.W. Allen. 2003. Semiconducting metal oxide sensor array for the selective detection of combustion gases. *Sens. Actuat. B* 93, 126–134.

Tonezzer, M., N.V. Hieu. 2012. Size-dependent response of single-nanowire gas sensors. *Sens. Actuators B Chem.* 163, 146–152.

Tricoli, A., M. Righettoni, A. Teleki. 2010. Semiconductor gas sensors: Dry synthesis and application. *Angew. Chem. Int. Edit.* 49, 7632–7659.

Wang, H.N., X.F. Zhou, M.H. Yu, Y.H. Wang, L. Han, J. Zhang, P. Yuan, G. Auchterlonie, J. Zou, C.Z. Yu. 2006. Supra-assembly of siliceous vesicles. *J. Am. Chem. Soc.* 128, 15992–15993.

Wang, H.N., Y.H. Wang, X. Zhou, L. Zhou, J. Tang, J. Lei, C.Z. Yu. 2007. Siliceous unilamellar vesicles and foams by using block-copolymer cooperative vesicle templating. *Adv. Func. Mater.* 17, 613–617.

Wang, J., H. Tang, H. Ren, R. Yu, J. Qi, D. Mao, H. Zhao, D. Wang. 2014. pH-regulated synthesis of multi-shelled manganese oxide hollow microspheres as supercapacitor electrodes using carbonaceous microspheres as templates. *Adv. Sci.* 1, 1400011.

Wang, J., M. Pan, J. Yuan, Q. Lin, X. Zhang, G. Liu, L. Zhu. 2020. Hollow mesoporous silica with a hierarchical shell from *in situ* synergistic soft–hard double templates. *Nanoscale* 12, 10863–10871.

Wang, Y., A. Pan, Q. Zhu, Z. Nie, Y. Zhang, Y. Tang, S. Liang, G. Cao. 2014. Facile synthesis of nanorod-assembled multi-shelled Co_3O_4 hollow microspheres for high-performance supercapacitors. *J. Power Sources* 272, 107–112.

Wang, Y., J. He, S. Yu, H. Chen. 2017. Solution-grown $CsPbBr_3/Cs_4PbBr_6$ perovskite nanocomposites: Toward temperature-insensitive optical gain. *Small* 13, 1701587.

Wang, Y., P. Liu, B. Zeng, L. Liu, J. Yang. 2018. Facile synthesis of ultralong and thin copper nanowires and its application to high-performance flexible transparent conductive electrodes. *Nanoscale Res. Lett.* 13, 78–84.

Wang, Z., H. Jia, Y. Cai, C. Li, X. Zheng, H. Liang, J. Qi, J. Cao, J. Feng, W. Fei. 2020. Highly conductive Mn_3O_4/MnS heterostructures building multi-shelled hollow microspheres for high-performance supercapacitors. *Chem. Eng. J.*, https://doi.org/10.1016/j.cej.2019.123890.

Wang, Y.R., S.H. Ma, L.F. Wang, Z.Y. Jiao. 2019. A novel highly selective and sensitive NH_3 gas sensor based on monolayer Hf_2CO_2. *Appl. Surf. Sci.* 492, 116–124.

Wanit, M., J. Yeo, S.J. Hong, Y.D. Suh, S.H. Ko, D. Lee. 2012. ZnO nano-tree growth study for high efficiency solar cell. *Energy Proced.* 14, 1093–1098.

Wiley, B., Y. Sun, B. Mayers, Y. Xia. 2005. Shape-controlled synthesis of metal nanostructures: The case of silver. *Chemistry* 11, 454–463.

Willimas, E., K.F.E. Pratt. 2000. Microstructure effects on the response of gas-sensitive resistors based on semiconducting oxides. *Sens. Actuators B Chem.* 70, 214–221.

Wong, Y.J., L. Zhu, W.S. Teo, Y.W. Tan, Y. Yang, C. Wang, H. Chen. 2011. Revisiting the Stöber method: Inhomogeneity in silica shells. *J. Am. Chem. Soc.* 133, 11422–11425.

Wu, H., L. Hu, M.W. Rowell, D. Kong, J.J. Cha, J.R. McDonough, J. Zhu, Y. Yang, M.D. McGehee, Y. Cui. 2010. Electrospun metal nanofiber webs as high-performance transparent electrode. *Nano Lett.* 10, 4242–4248.

Wu, H., D. Kong, Z. Ruan, P.C. Hsu, S. Wang, Z. Yu, T.J. Carney, L. Hu, S. Fan, Y. Cui. 2013. A transparent electrode based on a metal nanotrough network. *Nat. Nanotechnol.* 8, 421–425.

Wu, H., Y. Wang, C. Zheng, J. Zhu, G. Wu, X. Li. 2016. Multi-shelled NiO hollow spheres: Easy hydrothermal synthesis and lithium storage performances. *J. Alloys and Compounds* 685, 8–14.

Xiong, W., H. Liu, Y. Chen, M. Zheng, Y. Zhao, X. Kong, Y. Wang, X. Zhang, X. Kong, P. Wang, L. Jiang. 2016. Highly conductive, air-stable silver nanowire@iongel composite films toward flexible transparent electrodes. *Adv. Mater.* 28, 7167.

Xu, H., W. Wang. 2007. Template synthesis of multishelled Cu_2O hollow spheres with a single-crystalline shell wall. *Angew. Chem. Int. Ed.* 46, 1489–1492.

Xu, S.M., C.M. Hessel, H. Ren, R.B. Yu, Q. Jin, M. Yang, H.J. Zhao, D. Wang. 2014. α-Fe_2O_3 multi-shelled hollow microspheres for lithium ion battery anodes with superior capacity and charge retention. *Energ. Environ. Sci.* 7, 632–637.

Xu, W.H., L. Wang, Z. Guo, X. Chen, J. Liu, X.J. Huang. 2015. Copper nanowires as nanoscale interconnects: their stability, electrical transport, and mechanical properties. *ACS Nano.* 9, 241–250.

Xu, C.N., J. Tamaki, N. Miura, N. Yamazoe. 1991. Grain-size effects on gas sensitivity of porous SnO2-based elements. *Sens. Actuat. B* 3, 147–155.

Yamazoe, N., K. Shimanoe. 2009a. Receptor function and response of semiconductor gas sensor. *J. Sens.* 2009, doi: 10.1155/2009/875704.

Yamazoe, N., K. Shimanoe. 2009b. Receptor function of small semiconductor crystals with clean and electron-traps dispersed surfaces. *Thin Solid Films* 517, 6148–6155.

Yang, H.X., J.F. Qian, Z.X. Chen, X.P. Ai, Y.L. Cao. 2007. Multilayered nanocrystalline SnO_2 hollow microspheres synthesized by chemically induced self-assembly in the hydrothermal environment. *J. Phys. Chem. C* 111, 14067–14071.

Yang, X., X. Hu, Q. Wang, J. Xiong, H. Yang, X. Meng, L. Tan, L. Chen, Y. Chen. 2017. Large-scale stretchable semiembedded copper nanowire transparent conductive films by an electrospinning template. *ACS Appl. Mater. Interfaces* 9, 26468–26475.

Yang, Z., F. Xu, W. Zhang, Z. Mei, B. Pei, X. Zhu. 2014. Controllable preparation of multishelled NiO hollow nanospheres via layer-by-layer self-assembly for supercapacitor application. *J. Power Sources* 246, 24–31.

Ye, S., A.R. Rathmell, Z. Chen, I.E. Stewart, B.J. Wiley. 2014. Metal nanowire networks: the next generation of transparent conductors. *Adv. Mater.* 26, 6670–6687.

Yin, X.M., C.C. Li, M. Zhang, Q.Y. Hao, S. Liu, Q.H. Li, L.B. Chen, T.H. Wang. 2009. SnO_2 monolayer porous hollow spheres as a gas sensor. *Nanotechnology* 20, doi: 10.1088/0957- 4484/20/45/455503.

Yu, M.H., H.N. Wang, X.F. Zhou, P. Yuan, C.Z. Yu. 2007. One template synthesis of raspberry-like hierarchical siliceous hollow spheres. *J. Am. Chem. Soc.* 129, 14576–14577.

Yu, Y., F. Cui, J. Sun, P. Yang. 2016. Atomic structure of ultrathin gold nanowires. *Nano Lett.* 16, 3078–3084.

Yu, Y.T., P. Dutta. 2011. Examination of Au/SnO_2 core-shell architecture nanoparticle for low temperature gas sensing applications. *Sens. Actuat. B* 157, 444–449.

Zhan, K., R. Su, S. Bai, Z. Yu, N. Cheng, C. Wang, S. Xu, W. Liu, S. Guo, X.Z. Zhao. 2016. One-pot stirring-free synthesis of silver nanowires with tunable lengths and diameters *via* a Fe^{3+} & Cl^- co-mediated polyol method and their application as transparent conductive films. *Nanoscale* 8, 18121–18133.

Zhang, K., J. Li, Y. Fang, B. Luo, Y. Zhang, Y. Li, J. Zhou, B. Hu. 2018. Unraveling the solvent induced welding of silver nanowires for high performance flexible transparent electrodes. *Nanoscale* 10, 12981–12990.

Zhang, Y., J. Guo, D. Xu, Y. Sun, F. Yan. 2017. One-pot synthesis and purification of ultralong silver nanowires for flexible transparent conductive electrodes. *ACS Appl. Mater. Interfaces* 9, 25465–25473.

Zhang, J., S. Wang, Y. Wang, M. Xu, H. Xia, S. Zhang, W. Xia, X. Guo, S. Wu. 2009. ZnO hollow spheres: Preparation, characterization, and gas sensing properties. *Sens. Actuat. B Chem.* 139, 411–417.

Zhang, L.X., J.H. Zhao, J.F. Zheng, L. Li, Z.P. Zhu. 2011. Hydrothermal synthesis of hierarchical nanoparticle-decorated ZnO microdisks and the structure-enhanced acetylene sensing properties at high temperatures. *Sens. Actuat. B* 158, 144–150.

Zhang, T., L. Liu, Q. Qi, S.C. Li, G.Y. Lu. 2009. Development of microstructure In/Pd-doped SnO2 sensor for low-level CO detection. *Sens. Actuat. B* 139, 287–291.

Zhao, H., J.F. Chen, Y. Zhao, L. Jiang, J.W. Sun and J. Yun. 2008. Hierarchical assembly of multilayered hollow microspheres from an amphiphilic pharmaceutical molecule of azithromycin. *Adv. Mater.* 20, 3682–3686.

Zhao, J., M. Hu, Y. Liang, Q. Li, X. Zhang, Z. Wang. 2020. A room temperature sub-ppm NO$_2$ gas sensor based on WO$_3$ hollow spheres. *New J. Chem.* 2020, 5064–5070.

Zou, F., Y.M. Chen, K. Liu, Z. Yu, W. Liang, M.S. Bhaway, M. Gao, Y. Zhu. 2016. Metal organic frameworks derived hierarchical hollow NiO/Ni/graphene composites for lithium and sodium storage. *ACS Nano.* 10, 377–386.

Zuo, X., Y. Song, M. Zhen. 2020. Carbon-coated NiCo$_2$S$_4$ multi-shelled hollow microspheres with porous structures for high rate lithium ion battery applications. *Appl. Surf. Sci.* 500, 144000.

Chapter 3

Nanomaterials as Heavy Metal Sensors in Water

1. Introduction

In the current days, the major environmental threats to human beings are being introduced through heavy metal ions (HMIs) generation, which can be biodegraded easily (Aragay et al. 2011). Primarily, heavy metals are released in an anthropogenic way into the natural waters, which further cause a global epidemic through the release of wastewater, river water, tap water, soil, food and many organisms (Fu et al. 2011, Dehouck et al. 2016). These heavy metals have a severe toxic effect on the human body, by a generation of several serious diseases, viz., cancer, neurogenic disease and organic failure, which become very difficult to detect (Guo et al. 2017). These HMIs are constituted of metals and metalloids with a density of 4 g/cm^3, which is almost five times higher than that of water, and it may exist in the form of elements, ions and complexes (Ugulu et al. 2015). According to the United States Environmental Protection Agency (EPA), the list of toxic heavy metals include arsenic, mercury, nickel, copper, lead, cadmium and chromium. The presence of these metals in very low concentrations can also cause severe damage to the human body. Over the past few decades, plenty of detection methods have been developed in the process of detection of HMIs, viz., absorption spectrometry, fluorescence spectroscopy, electrochemical method, etc. (Mahmoud et al. 2018, Ping et al. 2011, Ping et al. 2012). The complex nature of spectroscopic studies restricts them from rapid detection on-site, and their high application cost creates a negative impact on the detection process. On the other hand, these conventional studies like atomic absorption spectroscopy (AAS), inductively coupled plasma/atomic emission spectrometry (ICP-AES), inductively coupled plasma-mass spectroscopy (ICP-MS), capillary electrophoresis (CE) and X-ray fluorescence spectroscopy (XFS) can easily detect the low concentrations of heavy metal ions, but they are highly sensitive and selective in nature. These days, chemical sensors have become very popular in on-site detection of multiple heavy metals, especially nanomaterials-based sensors create huge attention due to their large surface area, high catalytic efficiency, greater adsorption capacity and high surface reactivity. Mainly, carbon and silicon-based materials and metal nanoparticles have been utilized as potential sensor materials in heavy metal detection. This chapter will mainly emphasize the electrochemical sensors, fluorescent sensors, SERS sensors, plasmonic sensors, etc.

2. Electrochemical Sensors

The restriction of spectrometric methods for site purpose has been compensated largely by the application of effective electrochemical methods in the field of HMIs detection due to their portability, rapidity, simultaneous measurement and economic nature. Anodic stripping voltammetry (ASV) is the most commonly used technology in which deposition and stripping are usually applied. In recent days, disposable screen-printed electrodes (SPEs) have gained enormous popularity in comparison to the three-electrode testing system. In this method, three electrodes are integrated on an insulating substrate made of polyethylene terephthalate (PET), polyvinyl chloride (PVC) and ceramic materials. Therefore, the overall technology becomes portable, inexpensive and easy to operate (Barton et al. 2016). The SPE's thickness presents generally between a few microns to 100 μm, which can be in control with the various templates used and the amount of the ink used in printing purposes. Additionally, enhanced signal-to-noise ratio and improved mass transmission rate mainly can be attributed to the small electrode size, which allows the small volume of sample to detect. Especially, this technology does not involve any pre-treatment method like electrode polishing. On the other hand, the ink used for this purpose is mainly constituted of commercial conductive ink, such as silver ink, gold ink and carbon ink (Liu et al. 2019). Although, the non-conductive materials present in inks result in interference in the conductivity and electrochemical performance of the sensors.

In recent days, the inclusion of nanomaterials into SPE technology has become the most promising and effective step into the new era of the electrochemical detection method. The unique physical, chemical, optical and catalytic characteristics of the minute nanoparticles enhance the overall performance in various application fields, such as chemical/ biosensors, flexible electronics, energy devices, water purification, etc. (Zhao et al. 2018, Jiang et al. 2018, Choi et al. 2016). In electrochemical sensors, nanostructured materials with special characteristics, viz., high conductivity, wide electrochemical window and greater stability have gained huge attention. Nowadays, carbon nanomaterials in the form of carbon nanotubes and graphene, metal nano-particles, such as gold and silver, and metal oxides, e.g., Fe_3O_4 and Bi_2O_3 NPs, are usually taken into account.

(a) Carbon Nanomaterials

Carbon nanomaterials have become extremely popular as electrochemical sensors in the detection of heavy metals in the form of carbon nanotubes (CNTs), graphene (GR), graphene oxide (GO) and carbon nanofibers (CNPs).

(i) Carbon nanotubes were first introduced in the scientific community by Iijima in 1991. These special class nanomaterials consist of tubular shape with further categorization into two classes: single-walled carbon nanotubes (SWCNTs) and multi-walled nanotubes (MWCNTs). One of the most popular methods to synthesize SWCNTs is by arc-discharge method (Sarkar et al. 2018). Several research groups have recently reported the extensive application of SWCNT modified screen-printed sensors due to their high electrical conductivity, modulus, current density and stability (Luo et al. 2018, He et al. 2018). In a recent study, a SWCNT-enabled disposable sensor has been reported, which has been successfully utilized in stripping detection of Pb^{2+} ions (Molinero-Abad et al. 2018). In this work, the sensors are fabricated with drop-casting technology by the addition of the SWCNT solution onto the surface of the screen-printed electrode (SPE/SWCNTs). In the next step, gold nanoparticles get electrodeposited onto SPE/SWCNTs through the addition of $HAuCl_4$ solution. This sensor exhibits excellent performance with a linear range of 4–38 nM with a sensitivity of 72.128 nA nM^{-1} under the optimized conditions. This is analyzed through differential pulse anodic stripping voltammetry. This has also been applied to detect Pb^{2+} in seawater samples. On the other hand, MWCNTs implemented SPEs have successfully been applied in the detection of HMIs in the past few decades by owing two different routes. In the first option, MWCNTs are added to the printing ink. Injang et al. have adopted such a method in which disposable SPE doped with MWCNTs used in the detection of simultaneous determination of Cd^{2+}, Pb^{2+} and Zn^{2+} by addition of MWCNTs directly into the printing ink (Injang et al. 2010). Additionally, bismuth film is electroplated on the surface. In this process, the developed sensor shows a wide linear range of 2.0–100.0 mg L^{-1} for Pb^{2+}, 2.0–100.0 mg L^{-1} for Cd^{2+} and 12.0–100.0 mg L^{-1} for Zn^{2+} with the limit of detection (LOD) of 0.2 mg L^{-1} for Pb^{2+}, 0.8 mg L^{-1} for Cd^{2+} and 11.0 mg L^{-1} for Zn^{2+}. The other way to apply MWCNTs is to modify them on the surface of SPEs. Wang et al. developed a technology in which a portable screen-printed sensor is modified with MWCNTs onto the surface (Wang et al. 2017). This method is utilized to detect Cd^{2+} and Pb^{2+} simultaneously. The linear range for Cd^{2+} and Pb^{2+} is both from 1.0 to 60 mg L^{-1} and the LOD is 0.5 mg L^{-1} for Cd^{2+} and 0.12 mg L^{-1} for Pb^{2+}. This method is also successfully applied in a solid sample to detect HMIs.

(ii) Graphene was discovered in 2004, which became very popular in every aspect of the scientific community due to its high electron transfer rate, mechanical strength, electrical conductivity and large surface area value. In the field of sensing technology, graphene has taken an important position due to its wide electrochemical window, stability, large surface area and high electrical conductivity (Ping et al. 2011). Tang et al. develop a method in which graphene paste is directly printed onto a disposable screen-printed graphene electrode, which is used in rapid analysis of Cd^{2+} (Figure 3.1) (Teng et al. 2018).

The printing process is executed by passing graphene paste through a screen printed stencil (80 mesh) using a squeegee blade. From the electrochemical analysis, it is clear that screen-printed graphene electrode (SGPE) performs much better than those of graphite-enabled SPCE and glassy carbon electrodes, i.e., GCE. Under the optimized conditions,

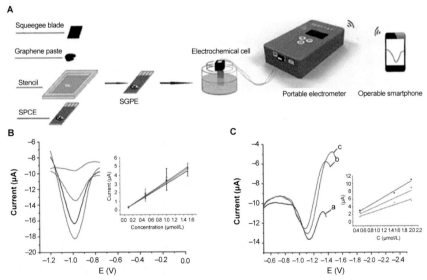

Figure 3.1: (A) Schematic illustration of the fabrication of SPGE using a screen-printing technique with GR paste for the analysis of trace Cd^{2+} (B) DPASV curves of different concentrations of Cd^{2+}, inset: standard addition curves of three rice samples. (C) DPASV curves of 0.5 (a), 1.5 (b) and 2.0 (c) mM Cd^{2+}, inset: standard addition curves of three rice samples (Teng et al. 2018) (Reproduced with permission from Teng et al. 2018).

the GR-enabled sensor exhibited a LOD as low as 10^{-7} M. Besides, the existence of calcium ion (Ca^{2+}), potassium ion (K^+), sodium ion (Na^+) and magnesium ion (Mg^{2+}) cannot interfere with the detection of Cd^{2+}. Moreover, the sensor was successfully employed for Cd^{2+} detection in rice samples. In another work, Shuai et al. reported SPGE technology to detect Cd^{2+} and Pb^{2+} simultaneously (Shuai et al. 2016). In this work, graphene is derived from graphene oxide, which is primarily oxidized from graphite powder. To further improve the viscosity and conductivity, n-butyl pyridinium hexafluorophosphate (IL) has been added to the graphene ink. This sensor shows enhanced activity with a wide linear range from 1.0 to 300.0 mg L^{-1} for both Cd^{2+} and Pd^{2+} with the same LOD as low as 0.1 µg L^{-1} for both metal ions. This sensor has been utilized in river water samples for Cd^{2+} and Pd^{2+} detection in a simultaneous manner. Another option to apply graphene in the sensor is to apply a fabricated graphene sheet embedded with an electrochemical sensor. Wu et al. developed graphene sheets by mixing the dispersed graphene oxide (GO) with hydrazine solution, then kept into stirring at 80°C for 24 hours (Wu et al. 2013). The graphene sheets then further dispersed into Nafion solution and dropped onto the surface of SPCE. To increase the sensing activity, Au nanoparticles electrodeposited onto the surface of SPCE/GR. The resultant sensing material exhibited a broad linear range of 0.5–60.0 mg L^{-1} for Pb^{2+} and 0.8–50.0 mg L^{-1} for Cd^{2+} with LOD value of 0.23 mg L^{-1} for Pb^{2+} and 0.35 mg L^{-1} for Cd^{2+}.

One of the major limitations in the application of graphene-based sensors is their toxic nature, therefore it can cause pollution with very ease. This particular reason restricts the researchers to apply potentially this type of sensor practically. The researchers are searching for effective green technology with enhanced sensing activity. Liu et al. reported a one-step electrodeposition technology to develop SPCE/GR where SPCE embedded directly in the graphene oxide suspension. Further, it was electrodeposited at –0.8 V for 600 s. In this typical green method, the toxic reducing agent has been avoided to apply (Liu et al. 2019).

(iii) Graphene oxides have also been utilized extensively and effectively in SPEs to detect HMIs. In typical research work, Ruengpiragiri et al. developed SPCE modified with GO to detect Cd^{2+}, Pd^{2+}, Cu^{2+} and Hg^{2+} (Ruengpirasiri et al. 2017). In this work, the classic Hummers' method had been utilized to synthesize GO (Hummers et al. 1958). In the next step, GO was mixed with carbon ink in the process of fabrication to establish GO doped SPCE (SPGOE). Here, antimony film was further deposited electrochemically to form SPGOE/SbF. The performance of this type of detector primarily depends on many key factors, viz., amount of GO in carbon ink, the concentration of Sb solution, supporting electrolyte and electrochemical deposition parameters. In this particular work, the sensor showed a linear response Cd^{2+}, Cu^{2+}, Pb^{2+} and Hg^{2+} at the same concentration scope of 0.1–1.5 mM and a low LOD of 54.0 nM for Cd^{2+}, 60.0 nM for Cu^{2+}, 26.0 nM for Pb^{2+} and 66.0 nM for Hg^{2+}. Finally, this sensor was employed for the detection of Cd^{2+}, Cu^{2+}, Pb^{2+} and Hg^{2+} in sewage, fertilizer waste and seawater samples. Another research group approached with a mixture of GO and diaminoterthiopene (GO-DTT) in simultaneous analysis of Cd^{2+}, Cu^{2+}, Pb^{2+} and Hg^{2+} with the LOD values of 1.9, 0.8, 2.8 and 2.6 µg L^{-1}, respectively (Choi et al. 2015).

(iv) There are many other types of carbon nanomaterials that have become very popular in the analysis process of HMIs, such as carbon nanofibers (CNFs), carbon nanohorns (CNHs) and carbon nanoparticles (CNPs). The size range of CNFs possesses between 0.5 to 100 µm in length. They are highly fibrous in nature and are usually synthesized by pilling graphite sheets in different patterns. Due to their compact structure, large surface area value and a small number of defects, they have proved themselves as one of the attractive choices in the field of HMIs detection through electrochemical sensing method (Perez-Rafols et al. 2017a, Perez-Rafols et al. 2017b). Carbon nanoparticles (CNP), a granular carbon nanomaterial with diameters ranging from dozens to hundreds of nanometers, become an effective material for electrode modification due to their large specific surface and remarkable stability (Yao et al. 2019, Niu et al. 2016).

(b) Metal Nanoparticles

With the unique physical and chemical characteristics of metal nanoparticles due to their very special structure, size and shape. Due to their novel and innovative properties, metal nanoparticles have been drawn enormous attraction in the field of energy, biomedical and sensors. With the excellent electrical conductivity and very high surface area value of this special class of materials, they have become the most potential candidates in the field of electrochemical sensors to detect heavy metal ions (Ahmad et al. 2018). In the last few decades, Au, Ag and Bi nanomaterials have shown the most effective performance in this research field. In the year 2010, Song et al. reported SPCE modified with Hg nano-droplets for the detection of Cd^{2+}, Pb^{2+} and Cu^{2+} (Song et al. 2010). In this case, MWCNTs-chitosan was used to coat the surface of SPCE. In this case, –1.3 V potential was applied. Finally, the LOD value had been evaluated as 12 nM for Cu^{2+} using the square wave anodic stripping voltammetry (SWASV) method. Another popular example of SPE sensors is Bi nanoparticles in HMIs detection. Rico et al. adopted a disposable electrochemical sensor in the process of detecting Zn^{2+}, Pb^{2+} and Cd^{2+} (Garando Rico et al. 2009). In this typical work, Bi nanoparticles were prepared using a one-step solution dispersion method using bismuth granules as the precursor materials. These nanoparticles were further coated on SPCE by a drop-casting technology. The LOD value was calculated as 2.6 mg L^{-1} for Zn^{2+}, 1.3 mg L^{-1} for Cd^{2+} and 0.9 mg L^{-1} for Pb^{2+} were obtained using the SWASV method. Similarly, Au and Ag nanoparticles were also extensively studied by a huge number of research workers. Niu et al. reported Au nanofilms which were potentially utilized in Hg^{2+} detection (Niu et al. 2012). Another research group had also developed SPCE modified AuNPs in the process of analysis of Zn^{2+}, Cu^{2+} and Pb^{2+} successfully (Lu et al. 2017). This sensor had been utilized for the detection of Hg^{2+} in human hair, urine and wastewater samples. Many other research groups had been reported on Ag nanoparticles which had been utilized successfully in HMEs through electrochemical studies (Domínguez-Renedo et al. 2007, Dominguez-Renedo et al. 2008, Vilian et al. 2017).

3. Optical Sensors

Optical chemical sensors (OCSs) have shown great potential in the on-site detection of HMIs using electromagnetic radiation by generating analytical signals in a transduction element. The interaction of the sample with radiation results in a transformation in the optical parameter, which can be directly related to the concentration of the analyte. The optical parameters include absorption, transmission, emission, lifetime, etc., through binding of an immobilized indicator, such as organic dye with the analyte sample (Gruber et al. 2017). The optical sensors are primarily categorized into three groups, fluorescent, surface-enhanced Raman scattering (SERS) and surface plasmon resonance (SPR).

(a) Fluorescent Sensors

This type of sensor functions through analyte-induced changes in the physicochemical properties of a fluorophore, which consider the intensity of fluorescence, life-time and anisotropy. These parameters are mostly related to the charge and energy transfer methods. A fluorescence nano-sensor consists of a fluorescent component that can generate signals toward the binding event and a receptor element for specific binding to the analyte (Buryakov et al. 2014). Quantum dots (QDs) have widely received popularity as an alternative to dyes based fluorophores as their excellent optical properties with a broad range of absorption spectra, greater quantum yield, narrow tuneable emission ability and high stability toward photobleaching. Therefore, greater photostability, outstanding water solubility and tuneable quantum size can greatly induce a wide emission range from UV to NIR prove them more suitable candidates in comparison to the conventional organic fluorescent dyes in biosensing applications (Li et al. 2013). The broad excitation profiles, narrow emission spectra, size-dependent fluorescence emission peaks, multiplexed detection capability, greater photostability and better quantum efficiencies make QDs more advantageous as fluorophore (Wang et al. 2017), which further can be divided into two groups, carbon dots (CDs) and graphene QDs (GQDs) as a potential candidate in HMIs detection with greater biocompatibility and attractive surface functionality (Ju et al. 2014, Li et al. 2014). Recently, an ion-imprinted dual emission QDs with the nano-hybrid structure of CdTe@SiO@CdSe based fluorescent device was developed for the detection of Cd^{2+} in water samples (Wang et al. 2016). Another research group had reported the synthesize of a novel luminescence solid-state material (QD-LDH) for the purpose of the on-site detection of Pb^{2+}, Cd^{2+} and Hg^{2+} ions in real water (Liu et al. 2017). This nano-sensor material had analyzed the samples with high selectivity and sensitivity with LOD of mixed metal ions 9.3×10^{-7} M. Chen et al. used hydrothermal preparation method to develop N-doped carbon (N-CQDs) as fluorescent prove to detect Hg^{2+} in the tap water sample and real lake water samples (Zhang et al. 2014). In another recent study, an Au nano-cluster with CdTe nano-composite had been developed to analyze Hg^{2+} and F^- ions. This probe was prepared from tripeptide-capped CdTe QDs and bovine serum albumin (BSA) protein-conjugated Au25 nanocluster (Paramanik et al. 2013). However, it is well known that CdS, CdSe and PbS contain toxic properties, whereas graphene oxides are non-toxic in nature and also possess carboxylic acid moiety. Therefore, these can be conjugated with amine-functionalized DNA and other bio-molecules by making covalent attachment. On the other hand, GOs are very economic in nature as fluorescence quencher, which can be easily synthesized in a large amount. This is the main reason, GOs have been considered as a promising alternative to QDs in HMI sensing. Many research groups are working extensively and effectively in the development of GO-based sensing materials. Chen et al. adopted a very easy synthesis method to produce high fluorescent N-doped GQDs in the detection process of Fe^{3+} (Ju et al. 2014). In another work, Lu et al. reported a one-step synthetic strategy of O- and S- co-doped graphitic carbon nitride QDs in which citric acid and thiourea had been used under heating conditions (Lu et al. 2015). These materials had shown a high quantum yield of 14.5% and also displayed different fluorescence responses in presence of various metal ions under the same conditions, although Hg^{2+} addition resulted in the quenching of the PL intensity of this special QD. This higher effective nature might be attributed to their strong affinity toward amino groups and thiourea groups present on the surface of the nano-materials with the

Hg^{2+} ions. Another important work had been reported by Kuo et al. who had developed a red-green-blue (RGB) switchable fluorescence resonance energy transfer (FRET) probe on the basis of fluorescent porous electrospun (ES) nano-fibers for effective detection of Hg^{2+} (Liang et al. 2017). In this typical process, ES nano-fibers were developed by electrospinning of poly (methyl methacrylate-co-1,8-napthalimide derivatives-co-rhodamine derivative) which consisted of naphthalimide derivatives along with a spirolactam rhodamine derivative. This particular FRET donor segment within the ES nanofibers showed high selectivity toward Hg^{2+} detection. In this process of detection, the green color transformed into blue color once it came into contact with Hg^{2+}. The color change phenomenon was attributed to the transformation of the thiourea unit to the imidazoline moiety, which resulted in the decrease of electron delocalization inside the fluorophore structures. In this case, the ES nano-fibers showed the specific selectivity in Hg^{2+} detection with the presence of other common ions, such as Zn^{2+}, Co^{2+}, Cd^{2+}, Pb^{2+}, Mg^{2+}, Ni^{2+}, Na^+, K^+, Cu^{2+}, Fe^{2+}, Mn^{2+}, Ca^{2+}, Pd^{2+} and Fe^{3+} (Zhu et al. 2015).

The mesoporous silica NPs (MSNPs) also had been considered as effective sensor materials due to their high stability and ease to fabricate for the abundant presence of hydroxyl groups onto their inner and outer surface. Additionally, their excellent solvent dispersibility made them attractive in HMI detection (Wan et al. 2013). In this regard, Sha et al. reported the development of fluorescent hybrid nanostructured materials by mesoporous silica nanoparticles with poly(p-phenylene vinylene) (PPV) to form PPV@MSN nano-composites (Sha et al. 2015). In this case, PPV can emit green light. In the next step, CdTe QDs had been fabricated onto the surface of MSNPs by using electrostatic interaction in which polyethyleneimine (PEI) has been worked as the polyelectrolyte. This material exhibited outstanding stability and dual-emission at 500 nm and 717 nm. This had also been applied as a fluorescent probe in the detection process of Cu^{2+} in water. This problem had shown excellent performance toward Cu^{2+} detection by changing the color from red to green with a LOD value of 31.2 nM.

(b) SERS Sensors

The conduction electrons present in plasmonic nanomaterials, such as AuNPs and AgNPs, the surface underwent through oscillations when excited by light of a specific wavelength, which led to a prominent electromagnetic field termed as LSPR. The LSPR is an extremely sensitive optical transducer. This depends on the aggregation state, size and shape of plasmonic nano-particles. Besides, the refractive index of the surrounding environment is also a key factor. SERS is a potential technology that can permit optically probing vibrational mode of individual bond for the ultrasensitive detection of analytes. In this case, the analytes get absorbed or interact with the metal nanostructures, which results in the enhancement of Raman scattering cross-section by several folds of magnitude due to SERS. SERS can also be utilized to detect single molecules. Moreover, it can differentiate different molecular orientations and highly similar molecules (Frost et al. 2015, Jeon et al. 2016). However, the reproducibility factor of SERS sensing nanostructures has been proved as poor, which had been attributed to their vulnerability of detachment from metallic nanoparticles. To reduce this drawback. Nanostructures had extensively been functionalized to gain interest. Zhao et al. had reported a combination of AuNPs with reduced GOs, which served as SERS substrates in the process of ultrasensitive and rapid detection of Pb^{2+} in water samples (Zhao et al. 2016). In this typical research study, $HAuCl_4$ had been reduced on GO nano-sheets to get AuNPs on rGO. This colloid had behaved as a flexible SERS platform with its regulated nano-architecture. This process further performed *in situ* chemical reactions of Pb^{2+} enhanced Au leaching. In another work, Frost et al. had adopted citrate functionalized AuNPs in the detection process of Pb^{2+} based on SERS (Liu et al. 2014). This work had detected Pb^{2+} with the LOD value of 25 ng L^{-1}. Chen et al. had also reported similar work with the application of AgNPs (Chen et al. 2014) with accurate detection of Hg^{2+} by LOD value of 0.34 nM.

(c) Plasmonic Sensors

In this type of sensor, the incident magnetic radiation can create free conduction electrons in noble metals, which can initially oscillate to result in SPR. In this case, the aqueous solution of 20 nm sized monodispersed AuNPs is primarily wine red, which in turn displays a strong SPR peak in the absorption spectrum. But the strong aggregation of AuNPs can cause transformation in color due to shifting of SPR peak. This particular principle is the basis of the development of colorimetric sensors in the detection process. Nie et al. reported a colorimetric sensor with Au nanoparticles of 15 nm diameter for potential performance with sensitive and specific detection of Hg^{2+}, Cu^{2+} and Ag^+ ions in water (Yang et al. 2017). Schopf et al. had also developed an optical sensor based on Au nanorods, which was immobilized onto a glass substrate for the detection of Hg by taking advantage of the plasmonic nature of Au and Hg (Schopf et al. 2017). In another important work, Ag-reduced GO nano-composites had been developed using *Moringa oleifera* fruit shell extracts as a reducing agent (Patil et al. 2017). Another research group had reported the formation of Diol-Au@Ag NPs in the detection of F^- ions (Kanyong et al. 2016). This nano-sensor exhibited high selectivity with a LOD value of 0.1 nM L^{-1}.

Table 3.1: Summary of HMIs determination by metal nano-particles enabled screen-printed electrochemical sensors.

Nanomaterials	Modification Method	Method	Analyte	Linear Range	LOD	Sample	Reference
AuNPs	Electrodeposition	LSV	Cr(VI)	20.0–200.0 µg L^{-1}	5.4 µg L^{-1}	River water	Tu et al. 2018
AuNPs	Electrodeposition	SWASV	Pb(II) Cu(II)	10–100 µg L^{-1} 10.0–150 µg L^{-1}	2.1 ng L^{-1}	Tap water	Kanyong et al. 2016
AuNPs	Drop-casting	LSASV	Pb(II)	0.4–20.0 mg L^{-1}	82.0 µg L^{-1}	River water, tap water	Shoub et al. 2017
AuNFs	Sputtering	SWASV	Hg(II)	1.2–8.0 µg L^{-1} 16.0–280.0 µg L^{-1}	0.8 µg L^{-1}	Unpolluted water, human hair, urine, wastewater	Niu et al. 2012
AgNPs	Electrodeposition	DPV	Cr(VI)	0.5–38.0 µM	0.85 µM	Tap water, sea water	Dominguez et al. 2007
AgNPs	Added in carbon ink	Potentiometry	Th(IV)	6.3–1.0 × 10⁸ nM	6.3 nM		Akl et al. 2016
AgNanoseeds	Drop-casting	ASV	Pb(II) Cu(II)	6.6–53.5 µg.L^{-1} 10.0–77.0 µg L^{-1}	1.98 µg L^{-1} 2.99 µg.L^{-1}	Ground water	Perez-Rafols et al. 2017
Ag nano-prisms	Drop-casting	ASV	Pb(II) Cu(II)	7.8–53.5 µg.L^{-1} 8.3–100.7 µg.L^{-1}	2.35 µg.L^{-1} 2.49 µg.L^{-1}	Ground water	Perez-Rafols et al. 2017
Ag@AuNPs	Drop-casting	SWV	Hg(II)	0.01–0.16 nM	6.0 pM	River water, tap water	Vilian et al. 2017
BiNPs	Electrodeposition	SWASV	Zn(II) Cd(II) Pb(II)	77.0–231.0 nM 44.0–134.0 nM 24.0–72.0 nM	8.0 nM 4.0 nM 2.0 nM		Malakhova et al. 2007

BiNPs	Drop-casting	SWASV	Zn(II) Cd(II) Pb(II)		2.6 µg.L^{-1} 1.3 µg.L^{-1} 0.9 µg.L^{-1}	Tap water	Granado Rico et al. 2009
MWCNTs-CHIT/Hg nano-droplets	Electrodeposition	SWASV	Cd(II) Pb(II) Cu(II)		12.0 nM 23.0 nM 12.0 nM	River water	Song et al. 2010
PtNPs	Electrodeposition	CV	As(III)	0.16–1.3 µM	0.077 µM	Tap water	Sanllorente-Mendez et al. 2009
SnNPs	Sparking deposition	SWASV	Cd(II) Zn(II)	1.0–30.0 µg.L^{-1}	0.5 µg.L^{-1} 0.3 µg.L^{-1}	Tap water, bottled water	Trachioti et al. 2018

References

Ahmad, R., O.S. Wolfbeis, Y.B. Hahn, H.N. Alshareef, L. Torsi, K.N. Salama. 2018. Deposition of nanomaterials: a crucial step in biosensor fabrication. *Mater. Today Commun.* 17, 289–321.

Akl, Z.F., T.A. Ali. 2016. Highly sensitive potentiometric sensors for thorium ions detection using morpholine derivative self-assembled on silver nanoparticles. *RSC Adv.* 6, 77854–77862.

Aragay, G., J. Pons, A. Merkoci. 2011. Recent trends in macro-, micro-, and nanomaterial-based tools and strategies for heavy-metal detection. *Chem. Rev.* 111, 3433–3458.

Barton, J., M.B.G. García, D.H. Santos, P. Fanjul-Bolado, A. Ribotti, M. McCaul, D. Diamond, P. Magni. 2016. Screen-printed electrodes for environmental monitoring of heavy metal ions: a review. *Microchim. Acta* 183, 503–517.

Buryakov, I.A., T.I. Buryakov, V.T. Matsaev. 2014. Optical chemical sensors for the detection of explosives and associated substances. *J. Anal. Chem.* 69, 616–631.

Chen, L., N. Qi, X. Wang, L. Chen, H. You, J. Li, C. Zhao, J.R. Lombardi. 2014. Ultrasensitive surface-enhanced Raman scattering nanosensor for mercury ion detection based on functionalized silver nanoparticles. *RSC Adv.* 4, 15055–15060.

Choi, C., M.K. Choi, T. Hyeon, D.H. Kim. 2016. Nanomaterial-based soft electronics for healthcare applications. *Chem. Nano. Mat.* 2, 1006–1017.

Choi, S.M., D.M. Kim, O.S. Jung, Y.B. Shim. 2015. A disposable chronocoulometric sensor for heavy metal ions using a diaminoterthiophene-modified electrode doped with graphene oxide. *Anal. Chim. Acta* 892, 77–84.

Clara, P.R., J.B. Arrieta, N. Serrano, J.M.M. Díaz-Cruz, C. Ariño, J.D. Pablo, M. Esteban. 2017. Ag nanoparticles drop-casting modification of screenprinted electrodes for the simultaneous voltammetric determination of Cu(II) and Pb(II). *Sensors* 17, 1458.

Dehouck, P., F. Cordeiro, J. Snell, B. de la Calle. 2016. State of the art in the determination of trace elements in seawater: a worldwide proficiency test. *Anal. Bioanal. Chem.* 408, 3223–3232.

Domínguez-Renedo, O., M.J. Arcos-Martínez. 2007. A novel method for the anodic stripping voltammetry determination of Sb(III) using silver nanoparticle modified screen-printed electrodes. *Electrochem. Commun.* 9, 820–826.

Domínguez-Renedo, O., L. Ruiz-Espelt, N. García-Astorgano, M.J. Arcos-Martínez. 2008. Electrochemical determination of chromium(VI) using metallic nanoparticle-modified carbon screen-printed electrodes. *Talanta* 76, 854–858.

Frost, M.S., M.J. Dempsey, D.E. Whitehead. 2015. Highly sensitive SERS detection of Pb2+ ions in aqueous media using citrate functionalised gold nanoparticles. *Sensors Actuators B Chem.* 221, 1003–1008.

Fu, F., Q. Wang. 2011. Removal of heavy metal ions from wastewaters: A review. *J. Environ. Manage.* 92, 407–418.

Granado Rico, M.A., M. Olivares-Marín, E. Pinilla Gil. 2009. Modification of carbon screen-printed electrodes by adsorption of chemically synthesized Bi nanoparticles for the voltammetric stripping detection of Zn(II), Cd(II) and Pb(II). *Talanta* 80, 631–635.

Gruber, P., M.P.C. Marques, N. Szita, T. Mayr. 2017. Integration and application of optical chemical sensors in microbioreactors. *Lab. Chip.* 17, 2693–2712.

Guo, J., Y. Kang, Y. Feng. 2017. Bioassessment of heavy metal toxicity and enhancement of heavy metal removal by sulfate-reducing bacteria in the presence of zero valent iron. *J. Environ. Manag.* 203, 278–285.

He, J., D. Yang, H. Li, X. Cao, L.P. Kang, X.X. He, R.B. Jiang, J. Sun, Z.B. Lei, Z.H. Liu. 2018. Mn_3O_4/RGO/SWCNT hybrid film for all-solid-state flexible supercapacitor with high energy density. *Electrochim. Acta* 283, 174–182.

Hummers, W.S., R.E. Offeman. 1958. Preparation of graphitic oxide. *J. Am. Chem. Soc.* 80, 1339–1339.

Injang, U., P. Noyrod, W. Siangproh, W. Dungchai, S. Motomizu, O. Chailapakul. 2010. Determination of trace heavy metals in herbs by sequential injection analysis anodic stripping voltammetry using screen-printed carbon nanotubes electrodes. *Anal. Chim. Acta* 668, 54–60.

Jeon, T.Y., D.J. Kim, S.-G. Park, S.-H. Kim, D.-H. Kim. 2016. Nanostructured plasmonic substrates for use as SERS sensors. *Nano Convergence* 3, 18.

Jiang, C.M., L.Y. Lan, Y. Yao, F.N. Zhao, J.F. Ping. 2018. Recent progress in application of nanomaterial-enabled biosensors for ochratoxin A detection. *TrAC Trends Anal. Chem.* 102, 236–249.

Ju, J., W. Chen. 2014. Synthesis of highly fluorescent nitrogen-doped graphene quantum dots for sensitive, label-free detection of Fe(III) in aqueous media. *Biosens. Bioelectron.* 58, 219–225.

Kanyong, P., S. Rawlinson, J. Davis. 2016. Gold nanoparticle modified screen-printed carbon arrays for the simultaneous electrochemical analysis of lead and copper in tap water. *Microchim. Acta* 183, 2361–2368.

Li, J., J.-J. Zhu. 2013. Quantum dots for fluorescent biosensing and bio-imaging applications. *Analyst* 138, 2506–2515.

Li, S., Y. Li, J. Cao, J. Zhu, L. Fan, X. Li. 2014. Sulfur-doped graphene quantum dots as a novel fluorescent probe for highly selective and sensitive detection of Fe^{3+}. *Anal. Chem.* 86, 10201–10207.

Liang, F.C., C.-C. Kuo, B.-Y. Chen, C.-J. Cho, C.-C. Hung, W.-C. Chen, R. Borsali. 2017. RGB-Switchable porous electrospun nanofiber chemoprobe-filter prepared from multifunctional copolymers for versatile sensing of ph and heavy metals. *ACS Appl. Mater. Interfaces* 9, 16381–16396.

Liu, J., G. Lv, W. Gu, Z. Li, A. Tang, L. Mei, W. Liu, W. Wang, L. Cao, M. Cheng. 2017. A novel luminescence probe based on layered double hydroxides loaded with quantum dots for simultaneous detection of heavy metal ions in water. *J. Mater. Chem. C* 5, 5024–5030.

Liu, M., Z. Wang, S. Zong, H. Chen, D. Zhu, L. Wu, G. Hu, Y. Cui. 2014. SERS detection and removal of mercury(ii)/silver(i) using oligonucleotide-functionalized core/shell magnetic silica sphere@au nanoparticles. *ACS Appl. Mater. Interfaces* 6, 7371–7379.

Liu, X., Y. Yao, Y. Ying, J. Ping. 2019. Recent advances in nanomaterial-enabled screen-printed electrochemical sensors for heavy metal detection. *Trn. Anal. Chem.* 115, 187–202.

Lu, Y.C., J. Chen, A.-J. Wang, N. Bao, J.-J. Feng, W. Wang, L. Shao. 2015. Facile synthesis of oxygen and sulfur co-doped graphitic carbon nitride fluorescent quantum dots and their application for mercury(II) detection and bioimaging. *J. Mater. Chem. C* 3, 73–78.

Lu, Z.W., J.J. Zhang, W.L. Dai, X.N. Lin, J.P. Ye, J.S. Ye. 2017. A screen-printed carbon electrode modified with a bismuth film and gold nanoparticles for simultaneous stripping voltammetric determination of Zn(II), Pb(II) and Cu(II). *Microchim. Acta* 184, 4731–4740.

Luo, B., H.L. Chen, Z.C. Zhu, B.J. Xie, C.S. Bian, Y.J. Wang. 2018. Printing single-walled carbon nanotube/nafion composites by direct writing techniques. *Mater. Des.* 155, 125–133.

Mahmoud, E., M. Ibrahim, N. Ali, H. Ali. 2018. Spectroscopic analyses to study the effect of biochar and compost on dry mass of canola and heavy metal immobilization in soil. *Commun. Soil Sci. Plant Anal.* 49, 1990–2001.

Molinero-Abad, B., D. Izquierdo, L. Pérez, I. Escudero, M.J. Arcos-Martínez. 2018. Comparison of backing materials of screen printed electrochemical sensors for direct determination of the sub-nanomolar concentration of lead in seawater. *Talanta* 182, 549–557.

Niu, P.F., C. Fernández-Sánchez, M. Gich, C. Navarro-Hernandez, P. Fanjul-Bolado, A. Roig. 2016. Screen-printed electrodes made of a bismuth nanoparticle porous carbon nanocomposite applied to the determination of heavy metal ions. *Microchim. Acta* 183, 617–623.

Niu, X.H., C. Chen, Y.J. Teng, H.L. Zhao, M.B. Lan. 2012. Novel screen-printed gold nano film electrode for trace mercury(II) determination using anodic stripping voltammetry. *Anal. Lett.* 45, 764–773.

Paramanik, B., S. Bhattacharyya, A. Patra. 2013. Detection of Hg^{2+} and F^- ions by using fluorescence switching of quantum dots in an Au-cluster-CdTe QD nanocomposite. *Chem. A Eur. J.* 19, 5980–5987.

Patil, P.O., P. V. Bhandari, P.K. Deshmukh, S.S. Mahale, A.G. Patil, H.R. Bafna, K.V. Patel, S.B. Bari. 2017. Green fabrication of graphene-based silver nanocomposites using agrowaste for sensing of heavy metals. *Res. Chem. Intermed.* 43, 3757–3773.

Perez-Rafols, C., J. Bastos-Arrieta, N. Serrano, J. Manuel Díaz-Cruz, C. Arino, J. de N.A. Malakhova, N.Y. Stozhko, K.Z. Brainina. 2007. Novel approach to bismuth modifying procedure for voltammetric thick film carbon containing electrodes. *Electrochem. Commun.* 9, 221–227.

Perez-Rafols, C., N. Serrano, J.M. Díaz-Cruz, C. Arino, M. Esteban. 2017a. Simultaneous determination of Tl(I) and In(III) using a voltammetric sensor array. *Sens. Actuators B* 245, 18–24.

Perez Rafols, C., N. Serrano, J. Manuel Díaz-Cruz, C. Arino, M. Esteban. 2017b. A screen-printed voltammetric electronic tongue for the analysis of complex mixtures of metal ions. *Sens. Actuators B* 250, 393–401.

Ping, J.F., J. Wu, Y.B. Ying, M.H. Wang, G. Liu, M. Zhang. 2011. Evaluation of trace heavy metal levels in soil samples using an ionic liquid modified carbon paste electrode. *J. Agric. Food Chem.* 59, 4418–4423.

Ping, J.F., J. Wu, Y.B. Ying. 2012. Determination of trace heavy metals in milk using an ionic liquid and bismuth oxide nanoparticles modified carbon paste electrode. *Chin. Sci. Bull.* 57, 1781–1787.

Ping, J.F., Y.X. Wang, K. Fan, J. Wu, Y.B. Ying. 2011. Direct electrochemical reduction of graphene oxide on ionic liquid doped screen-printed electrode and its electrochemical biosensing application. *Biosens. Bioelectron.* 28, 204–209.

Ruengpirasiri, P., E. Punrat, O. Chailapakul, S. Chuanuwatanakul. 2017. Graphene oxide-modified electrode coated with *in situ* antimony film for the simultaneous determination of heavy metals by sequential injection-anodic stripping voltammetry. *Electroanalysis* 29, 1022–1030.

Sanllorente-Mendez, S., O. Domínguez-Renedo, M.J. Arcos-Martínez. 2009. Determination of arsenic(III) using platinum nanoparticle-modified screen-printed carbon-based electrodes. *Electroanalysis* 21, 635–639.

Sarkar, T., S. Srinives. 2018. Single-walled carbon nanotubes-calixarene hybrid for sub-ppm detection of NO_2. *Microelectron. Eng.* 197, 28–32.

Schopf, C., A. Martín, D. Iacopino. 2017. Plasmonic detection of mercury via amalgam formation on surface-immobilized single Au nanorods. *Sci. Technol. Adv. Mater.* 18, 60–67.

Sha, J., C. Tong, H. Zhang, L. Feng, B. Liu, C. Lü. 2015. CdTe QDs functionalized mesoporous silica nanoparticles loaded with conjugated polymers: A facile sensing platform for cupric (II) ion detection in water through FRET. *Dye. Pigm.* 113, 102–109.

Shoub, S.A.B., N.A. Yusof, R. Hajian. 2017. Gold nanoparticles/ionophore-modified screen-printed electrode for detection of Pb(II) in river water using linear sweep anodic stripping voltammetry. *Sens. Mater.* 29, 555–565.

Shuai, H., Y.J. Lei. 2016. Graphene ink fabricated screen printed electrode for Cd^{2+} and Pd^{2+} determination in Xiangjiang river. *Int. J. Electrochem. Sci.* 11, 7430–7439.

Song, W., L. Zhang, L. Shi, D.W. Li, Y. Li, Y.T. Long. 2010. Simultaneous determination of cadmium(II), lead(II) and copper(II) by using a screen-printed electrode modified with mercury nano-droplets. *Microchim. Acta* 169, 321–326.

Teng, Y.J., Y.C. Zhang, K. Zhou, Z.X. Yu. 2018. Screen graphene-printed electrode for trace cadmium detection in rice samples combing with portable potentiostat. *Int. J. Electrochem. Sci.* 13, 6347–6357.

Trachioti, M.G., J. Hrbac, M.I. Prodromidis. 2018. Determination of Cd and Zn with "green" screen-printed electrodes modified with instantly prepared sparked tin nanoparticles. *Sens. Actuators B* 260, 1076–1083.

Tu, J.W., Y. Gan, T. Liang, H. Wan, P. Wang. 2018. A miniaturized electrochemical system for high sensitive determination of chromium(VI) by screen-printed carbon electrode with gold nanoparticles modification. *Sens. Actuators B* 272, 582–588.

Ugulu, I. 2015. Determination of heavy metal accumulation in plant samples by spectrometric techniques in Turkey. *Appl. Spectrosc. Rev.* 50, 113–151.

Vilian, A.T.E., A. Shahzad, J. Chung, S.R. Choe, W.S. Kim, Y.S. Huh, T. Yu, Y.K. Han. 2017. Square voltammetric sensing of mercury at very low working potential by using oligomer-functionalized Ag@Au core-shell nanoparticles. *Microchim. Acta* 184, 3547–3556.

Wan, X.J., S. Yao, H.Y. Liu, Y.W. Yao. 2013. Selective fluorescence sensing of Hg^{2+} and Zn^{2+} ions through dual independent channels based on the site-specific functionalization of mesoporous silica nanoparticles. *J. Mater. Chem.* 1, 10505–10512.

Wang, H., G. Zhao, Z.H. Zhang, Y. Yi, Z.Q. Wang, G. Liu. 2017. A portable electrochemical workstation using disposable screen-printed carbon electrode decorated with multiwall carbon nanotube-ionic liquid and bismuth film for Cd(II) and Pb(II) determination. *Int. J. Electrochem. Sci.* 12, 4702–4713.

Wang, J., C. Jiang, X. Wang, L. Wang, A. Chen, J. Hu, Z. Luo, S. Wang, S.S. Gambhir, S. Weiss. 2016. Fabrication of an "ion-imprinting" dual-emission quantum dot nanohybrid for selective fluorescence turn-on and ratiometric detection of cadmium ions. *Analyst* 141, 5886–5892.

Wang, Y.Y., X. Xiang, R. Yan, Y. Liu, F.-L. Jiang. 2017. Förster resonance energy transfer from quantum dots to Rhodamine B as mediated by a cationic surfactant: A thermodynamic perspective. *J. Phys. Chem. C*, Doi: 10.1021/acs.jpcc.7b08236.

Wu, F., H.W. Lin, X. Yang, D.Z. Chen. 2013. GS-Nafion-Au nanocomposite film modified SPCEs for simultaneous determination of trace Pb^{2+} and Cd^{2+} by DPSV. *Int. J. Electrochem. Sci.* 8, 7702–7712.

Yang, J., Y. Zhang, L. Zhang, H. Wang, J. Nie, Z. Qin, J. Li, W. Xiao, X.Y. Jiang, X.Y. Chen, Y. Zhang, H. Wang. 2017. Analyte-triggered autocatalytic amplification combined with gold nanoparticle probes for colorimetric detection of heavy-metal ions. *Chem. Commun.* 53, 7477–7480.

Yao, Y., H. Wu, J.F. Ping. 2019. Simultaneous determination of Cd(II) and Pb(II) ions in honey and milk samples using a single-walled carbon nanohorns modified screen-printed electrochemical sensor. *Food Chem.* 274, 8–15.

Zhang, J., L. He, P. Chen, C. Tian, J. Wang, B. Liu, C. Jiang. 2017. Silica-based SERS chip for rapid and ultrasensitive detection of fluoride ion by triggered cyclic boronate ester cleavage reaction. *Nanoscale* 9, 1599–1606.

Zhang, R., W. Chen. 2014. Nitrogen-doped carbon quantum dots: Facile synthesis and application as a "turn-off" fluorescent probe for detection of Hg2+ ions. *Biosens. Bioelectron.* 55, 83–90.

Zhao, F.N., J. Wu, Y.B. Ying, Y.X. She, J. Wang, J.F. Ping. 2018. Carbon nanomaterial enabled pesticide biosensors: design strategy, biosensing mechanism, and practical application. *TrAC Trends Anal. Chem.* 106, 62–83.

Zhao, L., W. Gu, C. Zhang, X. Shi, Y. Xian. 2016. *In situ* regulation nanoarchitecture of Au nanoparticles/reduced graphene oxide colloid for sensitive and selective SERS detection of lead ions. *J. Colloid Interface Sci.* 465, 279–285.

Zhu, C., G. Yang, H. Li, D. Du, Y. Lin. 2015. Electrochemical sensors and biosensors based on nanomaterials and nanostructures. *Anal. Chem.* 87, 230–249.

Chapter 4

Nanostructured Materials for the Development of Optical Fiber Sensors

1. Introduction

In recent days, optical fiber sensors generate a revolution in the field of biomedicine, environmental control, food quality testing and navigation systems (Culshaw et al. 2004). Some of the most relevant characteristics of optical fiber are its high bandwidth, lower transmission losses and the possible multiplex information. These sensors actually depend on different physical and chemical features and can detect multiple magnitudes, such as temperature, concentration, pressure, etc. (Cusano et al. 2008, Elosua et al. 2012). The key features of optical fiber sensors consist of their immunity to electromagnetic interferences, easy multiplexation, low weight and transmission losses, smaller size and real-time monitoring (Strobel et al. 2009, Jaroszewicz et al. 2005). For these advantageous reasons, optical sensing materials have been more popular over other electronic sensors in recent years. However, the electronic sensors have been developed with commercialization purposes in last few decades before the use of optical fiber was extended. Therefore, the prices are much lower. Additionally, the integrated application of electronic circuits and devices make them appealing and easy to manipulate. For this reason, it became essential to develop more potential optical fiber in an economical way.

In the development of a multi-sensor network, optical fibers allow remote sensing in favor of developing a multi-sensor network due to their low losses. In this case, no electrical biasing is needed to guide light, therefore it acts in a passive way. In the last few decades, thin-film coated optical waveguides have been popular in the researcher community

(James et al. 2006, Arregui et al. 2010). The performance was induced manyfold with the application of nanostructured thin-films and optical fibers (Sun et al. 2011, Arregui et al. 2009, Consales et al. 2012).

2. Fundamentals and Mechanisms of Optical Fiber Sensors

An optical fiber is a waveguide that can transmit light of various wavelengths. The propagation mechanism can be explained with Maxwell equations and also by ray theory as the wavelength of the signal is smaller than the physical dimensions of the waveguide. A standard optical fiber consists of silica, which can be classified into two parts: core and cladding. The core part should contain a slightly higher refractive index than the cladding's part. A small amount of energy always is transmitted from the core to the cladding part. In respect to the electromagnetic field, the signals mostly are transmitted by core modes; however, a small portion of it has been coupled into the cladding mode, which is termed as an evanescent field. Figure 4.1 presents the schematic diagram of the basic phenomenon. The transduction can happen with the core or cladding mode depending on the architecture and the sensing materials employed in the case of a sensor.

There are many ways to classify optical fibers. The optical fibers used in telecommunication systems are considered as standard by International Telecommunication Union (ITU), and it consists of 125 microns diameter. They can be classified into two groups according to the diameter: single-mode standard fiber (SMF) and multi-mode standard fiber (MMF). In the case of SMF, the diameter consists below 9 microns value, whereas MMF has a diameter value of 50 microns. In both the categories, core and cladding are made of SiO_2, but all the materials are made of silica. Some

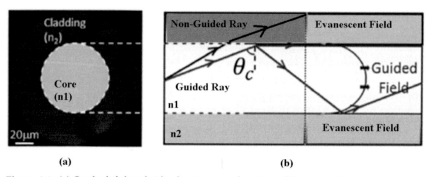

(a)　　　　　　　　　　　　　　(b)

Figure 4.1: (a) On the left-hand side, the transversal section of the optical fiber showing core (highlighted) and cladding; (b) on the right-hand side, the longitudinal section showing the signals involved in light propagation explained by ray theory (arrows) and electromagnetic field theories (reproduced with the permission from Elosua et al. 2017).

of them contain a silica core with plastic cladding, which are known as Plastic Cladding Silica (PCS) core fibers. This type of fiber has a diameter in the range of 100 to 1,100 microns. Additionally, the plastic cladding can remove easily, results in ease of interaction with the evanescent field. There is another type of fiber, which forms with plastic completely, named plastic or polymer optical fibers (POF). This type of fiber is successfully applied in sensors and biomedical field due to its mechanical robustness, excellent transmission properties and are easy to manipulate according to the application (Koike et al. 2011, Liehr et al. 2011, Bilro et al. 2012).

In the case of an optical fiber sensor, the transduction takes place when the target analyte transforms into the properties of light traveling through the waveguide. When the transduction happens externally to the optical fiber, the sensor becomes extrinsic. On the other hand, it is termed as intrinsic, when it occurs on the fiber. Mostly, all the sensors reported are intrinsic in nature. On the basis of the location of the sensing layer, two different configurations can be considered, viz., transmission and reflection. The sensing material is usually embedded into a supporting matrix material depositing at the end of a pigtail in the case of reflection sensors. The sensors developed with this structure are usually termed as optrodes as they look like typical electrodes. The key device consists of an optical coupler that can guide the signal from the light source to the sensor, but they yield low signal levels due to reduced interface. On the other hand, a direct optical path is created between the light source and receiver in the case of transmission sensors. Thus, the optical loss becomes lower in comparison to the other configurations. Although the resulting sensor is mechanically weaker, it needs to attach with a holder to prevent its damage. According to the applications, the transduction principle and the sensing material are usually taken into account.

3. Techniques for Development of Sensors

There are several ways of depositing sensing coatings onto optical fibers. The key advantage of this fiber is its small dimensions, but it becomes quite difficult to coat films in a uniform way. Thus, different techniques have been developed in handling the substrate.

(a) Layer-by-Layer (LBL) Nano Assembly

The main challenging factor in optical fiber synthesis is to deposit coatings and to control the morphology at a nano-metric level. LBL is such a process in which polyelectrolyte chains with different electrical charges get assembled to form micro-particles; the process was initiated in the 1960s (Iler et al. 1966). However, in the 90s, this technique became popular with

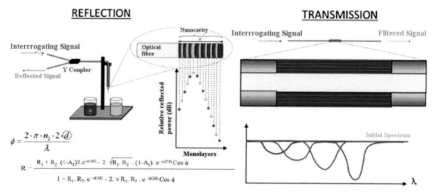

Figure 4.2: Mechanisms to control the nano-coatings growing for reflection and transmissions configurations: the first one is based on interferometry whereas the second on LMRs (reproduced with permission from Elosua et al. 2017).

its full potential for a wider application range (Decher et al. 1997). In this typical method, the substrate usually gets immersed into a polycationic or anionic solution for a certain amount of time to let the chains assemble. In between this time, rinsing is important as it removes the improperly assembled molecules. The morphology of the final products depends on the pH of the solutions, number of immersions, the concentration of polyelectrolyte and ionic strength. The researchers usually modify these parameters to obtain the desired morphology. This can be further used for both reflection and transmission architectures. The solutions are usually sprayed onto the substrates in the case of spray coating, which can form a thinner layer (Elosua et al. 2013). In the case of transmission configuration, rotation is needed to ensure homogeneous coating around the fiber. On the other hand, the solutions are sprayed in a perpendicular direction onto the end of the fiber in the case of transmission configuration.

(b) Dip Coating

Dip coating is a technique in which the fibers are dipping in vertical orientation into the solution or dispersion where the sensing materials are dissolved or dispersed and further fiber is removed from the solution. This is one of the easiest ways to execute the process. In this case, the key controlling factor is the withdrawal speed. The higher withdrawal speed leads to the formation of a thicker coating. The viscosity of the solution also plays an important role in this technique. In this case, the fiber is only immersed in a perpendicular direction, therefore this process is only applicable for reflection configuration or a hybrid one. If the material is coated at the end of the fiber, the film thickness can not be controlled. This method is typically applied in the deposition of sol-gel

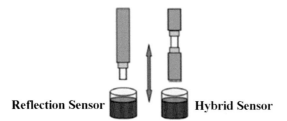

Reflection Sensor   **Hybrid Sensor**

Figure 4.3: Sensing configurations applicable for dip coating (reproduced with permission from Elosua et al. 2017).

and plastic matrices. A little variation can be applicable in the case of transmission sensors, which is termed as drop casting (González-Sierra et al. 2017).

(c) Electrospun Nano Webs

Electrospinning functions on the stretching of a viscoelastic solution formed through electrostatic forces; the materials are usually dissolved and the solvent employed is evaporated during the deposition process. This method is primarily performed with polymers and organic solvents (Wang et al. 2002). However, metal oxides and porphyrins are also successfully utilized in recent studies (Son et al. 2017, Urrutia et al. 2013). There are many parameters existing that can determine the morphology of the nano-fiber. In this case, if the solution is more viscous, a continuous jet would be ejected; on the other hand, if it is less, droplets would be formed. Other key factors are the solution conductivity, the applied electric tension, the distance between the electrodes, the flow rate of the syringe and the relative value of humidity. The electrode types can also control the performance, such as circular rotative discs and an array of counter electrodes can generate different alignments of the jet once it is deposited on the substrate. During the coating of nanoweb onto optical fibers, the substrate needs to be rotating. The rotating speed also will determine the morphology of the final supporting matrix.

4. Different Gas Sensing Probes

(a) Nano-Patterned Optical Fiber Tip in Gas Sensing

Nanostructuring is one of the most explored platforms in the area of nanophotonics as they possess light control properties at the nanoscale. Many research groups have worked on nano-patterning with the incorporation of metal or metal oxide to attain sub-wavelength structures for resonant field confinement and enhancement (Barnes et al. 2003,

Genov et al. 2004). The ability to form high-resolution nano-patterns on the micron-sized tip of optical fiber has potential applications in remote sensing. The implementation of fiber-tip-based sensors can offer huge advantages, specifically with respect to eliminating bulky optics and easy insertion of the device to the sensing area. The microscopic cross-sectional size and high mechanical flexibility of optical fibers allow *in vivo*, remote diagnostic detection. Few developments have been made wherein the fiber end facet has been changed to realize structures capable of detection of analyte molecules, such as localized surface plasmon resonance-based finger-tip probes have been realized for ultra-sensitive detection of protein biomarkers (Lin et al. 2011, Sanders et al. 2014). A finger-tip based Fabry-Perot (FP) has a sensing probe that consists of a silver layer and a vapor-sensitive polymer layer, which has been developed to detect methanol, acetone and hexanol vapors (Liu et al. 2009).

Due to the lack of development in suitable adsorption materials in detecting gas molecules, the sensitivity to surrounding gas concentrations remains a challenge. In recent days, graphene oxides (GO) have been utilized successfully to develop highly sensitive sensors due to their extraordinary high surface area to volume ratio (Robinson et al. 2008, Li et al. 2004). Moreover, plenty of functional groups are also present at the Go surface, and these groups can be used as traps for gas molecules (Nanda et al. 2016). However, GO and GO-based nano-composites have been applied in an extensive manner on conventional planar substrates in sensing gas species (Pandey et al. 2013). The integration of GO nano-sheets at the optical fiber tips to exploring sensing applications (Rosli et al. 2016). The standard fabrication technologies, such as electron beam lithography (EBL) (Consales et al. 2012), reactive ion etching (Jung et al. 2011) and liftoff (Sanders et al. 2014) have been utilized to pattern nanostructures on fiber facets. Another important technology has become very popular in recent days, viz., nano-imprint lithography (NIL) due to its excellent resolutions, simple and cost-effectiveness (Chou et al. 1995). An ultraviolet (UV) based NIL process has been reported for finger-tip patterning using non-trivial inclined UV incidence, making the process complicated and tedious (Scheerlinck et al. 2009). Another simple and efficient research work has been reported in which a high-resolution nanophotonic pattern has been inscribed on the cleaved facets using UV assisting technology (Viheriala et al. 2007).

(b) Nano-Crystalline ZnO in Optic Fiber Gas Sensing

Zinc Oxide (ZnO) has been considered as an interesting n-type semiconducting material with huge applications in the optoelectronic field and gas sensing devices. They possess large exciton binding energy

(60 meV), wide bandgap energy (3.37 eV) value. ZnO is chemically active and thermally stable with higher sensitivity to toxic and combustible gaseous materials. At the same time, it acts as a promising material in gas sensing applications due to its chemical sensitivity to volatile and other radical gases. It is quite abundant in nature, economic and non-toxic in nature (Wang et al. 2004). In recent days, metal oxide semiconductor materials are drawing huge attention as gas sensors due to their exceptional nature in rapid response and recovery in the detection of polluted and toxic gaseous components (Zhang et al. 2009, Kim et al. 2011). ZnO, SnO_2, TiO_2 and WO_3 thin films have been explored extensively as promising gas sensors by researchers (Xu et al. 2008, Kim et al. 2010, Moon et al. 2011, Xiang et al. 2010). The applications of these metal oxides are nanostructured as fiber optic sensing probes have more advantageous conditions due to their lighter weight, smaller size and higher sensitivity (Lee et al. 2003). In another work, Valerini et al. proposed nanostructured ZnO films with different compositions that had been utilized in NO_2 gas sensing (Valerini et al. 2010).

(c) Carbon Nano-Materials in Fiber-Optic Gas Sensors

Sensors with optical components as architectural substrates have drawn huge attention across the scientific community due to their numerous analytical advantageous features in sensing applications encompassing different chemicals, gaseous materials and biomolecules. Optical gas sensing probes usually rely on the phenomenon of specific modification in the optical characteristics of the sensing surface coated over the optical substrate, such as a prism, grating or optical fiber, on exposure to target gas molecules in a specific spectral range that is manifested in respect to the measurable and decipherable optical signal. Carbon nanotubes and graphene have become the most extensively employed materials in gas sensing applications due to their extraordinary electrical and optical properties, which projected them as a potential choice in synthesizing gas-sensitive surfaces. The multi-walled carbon nanotubes (MWCNTs) possess a large surface area and excellent electric conductivity due to their high electron mobility, which leads to efficient interaction and effective absorption of gaseous species on the sensing surfaces (Llobet et al. 2013, Angiola et al. 2016). On the other hand, graphene, pioneering 2-D carbon nanomaterials consist of greater specific surface area, zero bandgap and high electrical conductivity value along with low Johnson noise (Yao et al. 2014, Vikrant et al. 2018). With exposure to the target gas molecules, those get absorbed on the graphene incorporated sensing surface and behave as either electron acceptor or donor entities. The interaction between them changes in an appreciable manner with the variation in the carrier concentration of graphene, which results in the modification of its electrical and optical properties. These properties finally generate variation in the optical spectrum of the sensor. The excellent gas sensing properties of these nanomaterials are attributed to functionalization with oxygen groups that generate oxygen vacancies on the sensing surfaces formed using CNTs and graphene. There are numerous research groups that have reported the application of carbon nanomaterials in gas sensing probes (Llobet et al. 2013, Angiola et al. 2016).

5. Gas Sensing

(a) Ammonia

Ammonia is a colorless gas with a pungent smell, which is extensively utilized as a chief component in the industrial manufacturing of various fertilizers and explosives. Moreover, NH_3 is consumed as a raw product and produced as by-products in various industrial operations. The threshold level of NH_3 as per the industrial standards is 25 ppm, inhalation beyond which may result in acute poisoning and detrimental to human health. Various fiber-optic sensors for NH_3 with sensing surfaces with metal oxides and carbon nanomaterials have become very popular these days among the scientific community. Especially, ZnO, TiO_2, SnO_2 and ITO have been extensively utilized in the fabrication of NH_3 sensors due to their extraordinary sensitivity toward NH_3 molecules. Renganathan et al. reported a fiber-optic NH_3 sensor in which nano-crystalline ZnO was adopted which was annealed with different temperatures, such as 500° and $1,200^\circ C$ (Renganathan et al. 2011). The interaction of NH_3 gas molecules with ZnO leads to a modification of optical properties, results in the sensing mechanism. The sensing probe was analyzed for NH_3 gas concentration in the range 0–500 ppm in which a linear trend in the intensity of the spectral peaks was observed with the ZnO layer annealed with $1,200^\circ C$ turning up to be more sensitive in comparison to that with $500^\circ C$. They had also claimed that the selectivity of the sensor in presence of methanol and ethanol vapors and have reported a response time of 100 minutes along with 80 minutes recovery time. On the other hand, Bhatia and Gupta had reported the sensitivity of NH_3 molecules toward dyes, such as bromocresol purple (BCP) which had been exploited to fabricate a fiber-optic SPR based NH_3 sensor (Bhatia et al. 2013). They had executed the work with a 70 nm thick BCP layer that was used as a SPR framework for NH_3 sensing deposited over consecutive layers of silver (40 nm) and silicon (5 nm) coated over a barecore of a silica optical fiber. The SPR characterization of the fabricated fiber-optic probe was examined in a wavelength interrogation scheme using NH_3 gas concentration in the range of 10–150 ppm and a red shift equal to 30 nm in resonance wavelength was shown. The sensing mechanism involved the formation of a complex between NH_3 and BCP was attributed to the diffusion of NH_3 molecules into the BCP layer on the exposure of BCPcoated sensing probe to NH_3 gas in continuation with the below-mentioned chemical equation:

$$BCP + NH_3 + H_2O \leftrightarrows BCPOH^-. NH_4^+$$

The formation of the complex mentioned in the equation brings about arise in the dielectric constant value of the BCP layer resulting in a red shift in resonance wavelength. They had further demonstrated the tunability of the operating range of the sensors with variation in silicon thickness inside the sensor configuration. Mishra et al. had adapted an SPR based fiber-optic sensor employing ITO as material for generating surface plasmons to detect NH_3 in a low concentration range of 1–10 ppm (Mishra et al. 2015). In another report, Mishra et al. developed a nanocomposite arrangement of poly (methyl methacrylate) (PMMA) and reduced graphene oxide (rGO) in ammonia sensing which was deposited over copper-coated unclad core region of an optical fiber (Mishra et al. 2014). The PMMA/rGO nanocomposite was synthesized by an *in situ* bulk polymerization process with various wt.% of rGO ranges from 0.5 to 15. The sensors were analyzed with SPR measurements for NH_3 gas concentration in the range of 10–100 ppm via which the researchers showed that the sensing probe having 5 wt.% of rGO in PMMA/rGO nanocomposite. These materials had shown a maximum shift in resonance wavelength amounting to 35 nm in comparison with those fabricated with other values of rGO wt.%. Another group had studied the influence of the thickness of the sensing layer used for NH_3 detection with a theoretical analysis of the electric field intensity at sensing surface-gas analyte (NH_3) interface. In this case, tin oxide (SnO_2) was used as the sensor transducer layer whose thickness was optimized by using a numerical measurement of electric field intensity at every interface in the fiber-optic sensors. This probe was configured as unclad silica fiber core/Ag/SnO_2. In this typical work, the authors firstly optimized the thickness of the SnO_2 layer experimentally through SPR studies which were further deduced to be 18 nm in attaining maximum sensitivity, using which the SPR characterization of the fiber-optic probe was carried out using NH_3 gas in the concentration range 10–100 ppm. After that, they executed a numerical simulation method to study the variation of electric field intensity at each interface of the fabricated sensing probe with an optimum value of SnO_2 thickness 24–34 nm. Raj et al. reported a taper-shaped fiber-optic probe coated with silver nanoparticles/PVP/PVA hybrid as the sensing surface (Mishra et al. 2015). A summary of the performance parameters of different reports utilizing different sensing layers over the unclad fiber core is presented in Table 4.1.

Table 4.1: Summary of performance parameters of some of the significant reports for NH$_3$ detection utilizing different sensing layers coated over unclad optical fiber core.

Sensing Layer	Sensitivity	LOD (ppm)	Calibration Range (ppm)	Response Time	Reference
ITO/Pani	–	–	10–150	–	Mishra et al. 2012
Nanocrystalline ZnO	–	–	0–500	–	Renganathan et al. 2011
ZnO NPs incorporated GO	–	–	4–140	–	Fu et al. 2018
Ag/Si/BCP	0.45 nm/ppm	0.733	10–150	–	Bhatia et al. 2013
ITO/BCP	1.891 nm/ppm	0.175	1–10	–	Mishra et al. 2015
Cu/PMMA/rGO nanocomposite	0.9 nm/ppm	0.367	10–100	–	Mishra et al. 2014
Ag/SnO$_2$	2.15 nm/ppm	0.154	10–100	–	Pathak et al. 2015
Ag NPs/PVP/PVA hybrid	0.88 counts/ppm	–	0–500	–	Raj et al. 2015
Polyphyrin incorporated TiO$_2$	–	0.1	0.1–1 (gaseous) 5–100 (solution)	< 30s	Tiwari et al. 2017
Ag nanowires	0.17 counts/ppm	–	0–500	70 min	Shobin et al. 2014
PDDA functionalized TSPP	–	0.5	0.5–50	< 3 min	Korposh et al. 2018
Pt NPs incorporated GO	10.2 pm/ppm	–	0–120	–	Yu et al. 2017
NiO nanocrystals	–	–	0–500	–	Yamini et al. 2017
Electrospun PPO nanofibers	–	0.1	0.1–10	–	Bagchi et al. 2017
SWCNTs	–	–	0–500	60 min	Manivannan et al. 2012

(b) Chlorine

Chlorine is a gas used as a disinfectant and oxidizing agent in different research and industrial operations besides imparting a significant role in bleaching methods and pharmaceutical drug formulations. Chlorine has been extensively used in water purification; however, the consumption of chlorinated water beyond a particular value can cause a lot of disease in the human body. On the other hand, the inhalation of higher concentrations of Cl_2 gas by human beings can result in the irritation of skin, eyes and lungs to a crucial extent. Mishra and Gupta had reported applying a nanolayer indium oxide (In_2O_3) doped tin oxide (SnO_2) as a chlorine sensing surface, which was deposited over Ag coated 1 cm unclad core of optical fiber (Mishra and Gupta 2015). They had characterized the fabricated fiber-optic Cl_2 gas sensing probe by SPR studies carried out in wavelength interrogation module using Cl_2 concentration in the range 10–100 ppm and a red shift of 14 nm in the resonance wavelength that had been reported. This sensor had been revealed to be more sensitive in the detection of low concentration of Cl_2 gas. The atomic weight percent of In_2O_3 and SnO_2 with a thickness of sensing layer thickness was optimized to 6 at wt% and 12 nm, respectively. In this optimized condition, the maximum shift in resonance wavelength was reported. Moreover, the significance of In_2O_3 doping over SnO_2 had also been reported. The resonance wavelength shifted more with the doping of In_2O_3 over SnO_2 in comparison to those of only In_2O_3 and SnO_2 layers over the silver-coated sensing region of the probe. The doping of In_2O_3 actually increased the defect levels in the sensing surface, which further facilitated the formation of an excess number of active sites for Cl_2 molecules to interact with the medium. The same group had also reported one work in which ZnO had been utilized in the fabrication of SPR based fiber-optic Cl_2 sensor (Usha et al. 2015). In this work, an 18 nm thick ZnO layer was deposited over a 40 nm thick silver layer over the unclad silica core of an optical fiber with a maximum shift in resonance wavelength and maximum sensitivity. They had also analyzed SPR characterization of the probe at 10–100 ppm concentration of Cl_2 gas and demonstrated the sensitivity of the sensor under the influence of different gases, such as H_2S, NH_3, CH_4, H_2 and N_2. The same research group had developed another SPR based sensor to monitor chlorine content dissolved in water, which consisted of a 5 nm thick layer of conducting polymer polyvinylpyrrolidone (PVP) as the sensing element deposited over consecutive layers of silver (40 nm) and ZnO (10 nm) coated unclad core of an optical fiber (Tabassum et al. 2015). In this typical process, the refractive index of PVP had been changed with exposure to solutions enriched with Cl^- ions by following the below-mentioned chemical reaction:

$$H_2O \text{ (l)} + Cl_2 \text{ (g)} \leftrightarrows HClO \text{ (l)} + HCl \text{ (l)}$$

By following the reaction, a detectable SPR signal had been generated. They had examined the SPR behavior of the fiber-optic probe for Cl⁻ contained solutions with 0.5–5 ppm concentration as a blank solution for the reference purpose, which resulted in a red shift of 12 nm in the resonance wavelength as elucidated by SPR spectra. In the optimization of PVP layer thickness, it had been revealed that 5 nm thick PVP layer achieved the maximum shift in resonance wavelength.

(c) Hydrogen

It is an element that had achieved appreciable attention both in gaseous and liquid states. Moreover, it has drawn maximum popularity as an efficient source of clean and renewable energy source with ample abundance, thus establishing this element as the most powerful and promising candidate in alternative to the fossil fuels sources. Due to its low atomic weight and high diffusion coefficient value of 0.16 cm²/sec in air, H_2 can easily migrate from one place to another, which put forth a challenge toward preventing hydrogen gas leakage. Additionally, it is a highly flammable and explosive gas as its ignition energy value is very low (0.018 mJ) and has a high combustibility value of 285.8 KJ/mol. It has also wide chances of getting exploded in the air, which can be between 4 to 75% volumetric concentration. These above-mentioned facts make this gas a serious threat to storage at a mass-scale level and also creates a problem in the commercial deployment of hydrogen gas. Thus, a robust and highly sensitive sensing probe becomes essential with effective specifications, which could be fulfilled by fiber-optic SPR sensing probes. Many research groups had reported the potential sensitivity of hydrogen toward palladium (Pd), which had been successfully applied in fiber-optic H_2 sensors. It was mentioned in the literature that hydrogen on the Pd surface results in a phase change in its crystallography (from α to β) at room temperature and normal atmospheric pressure. This phenomenon leads to the formation of a reversible palladium hydride (PdH$_x$; in which x stands for the atomic ratio of H/Pd) by following the reaction:

(α)Pd + x/2 H$_2$ ⇋ (β)PdH$_x$

The absorption property was enhanced 900 times its volume of H_2 gas due to this phase change phenomenon, which resulted in a transformation in the complex dielectric function of palladium according to the equation:

$$\xi_{Pd}(c) = h(c) \times \xi_{Pd}(0)$$

In which $\xi_{Pd}(c)$ and $\xi_{Pd}(0)$ represent the complex dielectric function value of Pd at two different conditions, (i) in presence of hydrogen and

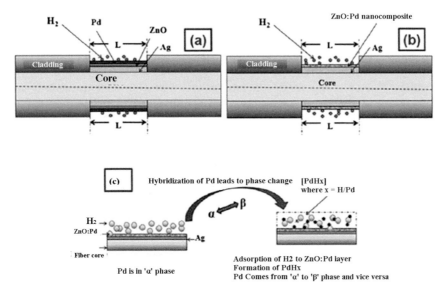

Figure 4.4: (a) Schematic illustration of SPR based fiber-optic probe for H_2 detection having Ag/ZnO/Pd multilayers, (b) SPR based fiber-optic H_2 sensing probe having Ag/ZnO:Pd nanocomposite arrangement, (c) illustration of sensing mechanism, (d) variation of FOM as a function of the thickness of ZnO layer for the multi-layered sensing probe Ag/ZnO/Pd for varying Pd thickness (reproduced with permission from Tabassum et al. 2015).

(ii) in absence of hydrogen. The h(c) is a non-linear function with values of h(0) = 1.0 and h(4) = 0.8 in respect to 0 and 4% concentrations of H_2 gas. In accordance with this, Tabassum and Gupta investigated a theoretical study of a SPR based fiber-optic hydrogen sensor with thin films of ZnO and Pd in a multi-layered arrangement as well as ZnO:Pd nanocomposite, which was deposited over Ag coated unclad silica fiber core (Tabassum et al. 2015). Figure 4.4(a) and (b) represented the schematic diagram of the H2 sensor having the multi-layered configuration of ZnO and Pd, and their nanocomposite arrangement with ZnO:Pd, respectively, over Ag coated unclad core of the optical fiber; whereas Figure 4.4(c) shows the sensing mechanism.

This research group analyzed through numerical simulations that the probe with ZnO:Pd nanocomposite layer attained a much greater shift in resonance wavelength of 0.139 μm in the case of ZnO:Pd with a thickness of 0.080 μm and volume fraction of 0.55 with Pd in ZnO, which had been compared with that of ZnO and Pd multi-layered structure of 0.020 μm for ZnO and Pd thickness value of 0.090 μm and 0.050 μm, respectively. This theoretical model had been executed experimentally in their other work already reported above (Tabassum et al. 2015).

However, the phenomena of development of PdHx might give rise to certain issues that further concerns the durability of the fiber-optic sensor due to the enhanced mechanical stress on the Pd lattice. These issues can be rectified by alloying Pd with another suitable metal which additionally could enhance the hydrogen permeability via the sensing layer that can result in the reduction in stress at Pd lattice. This will enhance the overall life of the hydrogen sensing probe. One research group had incorporated PdY as a hydrogen sensing probe (Liu et al. 2012). Although this resulted in the employment of a more thick Pd layer, which restricted the response time to some extent. This restriction was further addressed by another research group, who had developed silver, SiO_2 and PdY alloy in multi-layered consequence over an unclad core of a silica optical fiber to numerically designed hydrogen sensing probe (Downes et al. 2017). In this case, the silver layer acted as SPR active material, whereas SiO_2 functioned as a modulation layer that coupled the guided rays in fiber core with surface plasmons at the core-metal interface and simultaneously increased the electric field intensity at fiber-core metal interface. On the other hand, PdY was acted as a hydrogen sensitive layer. The authors investigated the sensor response via the theoretical method and calculated the optimum values of silver and SiO_2 layers as 50.5 and 72 nm, respectively. The proposed sensing probe had exhibited a sensitivity of 17.64 nm in correspondence to 4% H_2 gas concentration. Another research group had reported the development of thin films of ITO as the sensing probe for SPR based fiber-optic H_2 sensor which was functioned in an intensity interrogation scheme (Mishra et al 2012). In this typical work, the ITO layer was developed using various composition ratios of In_2O_3 and SnO_2 in 90:10, 70:30, 50:50 and 30:70. The reactivity of ITO with hydrogen resulted in a change in the dielectric constant value of the ITO layer; this phenomenon further had led to the change in resonance wavelength. Two major factors had been taken into consideration, the relative composition of In_2O_3 and SnO_2 and the thickness of their corresponding layers. It had been noticed that the maximum shift in resonance wavelength occurred with the composition ratio of 70:30 in $In_2O_3:SnO_2$ and with 90 nm ITO thickness. This sensing probe recorded SPR spectra corresponding to 100% N_2 and (4% H_2 and 96% N_2) gases at different time intervals.

In another work, Ta_2O_5 and Pd coated uniformly over a hetero core-structured optical fiber was exhibited a cylindrical geometry without removing the fiber cladding (Hosoki et al. 2013). The researchers evaluated an SPR Kretschmann configuration based fiber-optic sensor by insertion of 15 mm long portion of a single-mode optical fiber constituting core and cladding diameter of 3 μm and 125 μm, respectively, which turned into a multi-mode fiber with core diameter 50 μm and cladding diameter identical to that of single-mode fiber. They had demonstrated a 28 nm shift in resonance wavelength range with respect to 0 and 4% concentrations of H_2 gas in nitrogen under the optimized thickness of different layers with Ta_2O_5 layer of 60 nm thick and Pd layer of 5 nm thick, which had been embedded over 25 nm thick gold layer. In another work, Wang et al. had developed a fiber-optic SPR hydrogen sensing probe using a nano-composite arrangement of Pt/WO_3 as the sensing layer (Wang et al. 2013). In this case, a SiO_2 layer was placed in between the layers of plasmonic metal silver and the sensing surface, which acted as a modulation layer. This was coated consequently over a step-index optical fiber of core diameter 200 μm and N.A. 0.22. During exposure of hydrogen gas over Pt/WO_3 nano-composite layer, the dissociation of hydrogen molecules and the subsequent adsorption of hydrogen atoms intercalated into WO3 layer via the spillover process. Due to the chemical reaction that happened over Pt/WO3 layer, the dielectric function value had been changed, which further resulted in a shift attained in resonance wavelength and finally developed the sensor working principle.

$$H_2 \xrightarrow{Pt} 2H_{ad} \xrightarrow{Spillover} 2H^+ + 2e^-$$

$$WO_3 + xH^+ + xe^- \rightarrow H_x WO_3$$

In this case, a resonance wavelength shift had been evaluated as 17.4 nm for the optimized sensor Ag of 35 nm thickness, SiO_2 of 100 nm, WO_3 of 180 nm and Pt of 3 nm in respect to 0 and 2% concentration of hydrogen gas in an Ar atmosphere.

(d) Methane

Methane is a commonly used gas for cooking purposes. Additionally, it served in a lot of industrial applications pertaining to the production of numerous chemicals, fertilizers and pharmaceuticals. CH4 is a highly flammable gas and is very prone to explosion. Therefore, its timely detection is highly required. The fabricated SPR based fiber-optic methane gas sensing probe using multi-layers of Ag and graphene-CNTs/poly(methyle methacrylate) hybrid composite had been developed (Mishra et al. 2015). In this case, the adsorption of CH4 molecules on the surface transformed the dielectric constant value. The unique properties of GCNT/PMMA composite had been evaluated as high aspect ratio and tailored optical properties. SPR analysis on the sensor using 10–100 ppm concentration of methane gas resulted in a shift of 30 nm in resonance wavelength for 5 wt% doping concentration of GCNT in PMMA matrix. They had also compared the shift with the sensing probe consisting of only GCNT, CNT and rGO layers over Ag coated unclad fiber core. Additionally, optimization of wt% of GCNT in PMMA was performed as doping of GCNT in PMMA, which resulted in the enhancement of defect levels. Thus, the active sites situated in the PMMA matrix facilitated efficient gas adsorption.

References

Angiola, M., C. Rutherglen, K. Galatsis, A. Martucci. 2016. Transparent carbon nanotube film as sensitive material for surface plasmon resonance based optical sensors. *Sens. Actuators B* 236, 1098–1103.

Arregui, F.J., I.R. Matias, J. Goicoechea, I. Del Villar. 2009. Optical fiber sensors based on nanostructured coatings. pp. 275–301. *In*: Arregui, F.J. (ed.). Sensors based on Nanostructured Materials. New York: Springer.

Arregui, F.J., I.R. Matias, J.M. Corres. 2010. Optical fiber sensors based on Layer-by-Layer nanostructured films. *Procedia Eng.* 5, 1087–1090.

Bagchi, S., R. Achla, S.K. Mondal. 2017. Electrospunpolypyrrole-polyethylene oxide coated optical fiber sensor probe for detection of volatile compounds. *Sens. Actuators B* 250, 52–60.

Barnes, W.L., A. Dereux, T.W. Ebbesen. 2003. Surface plasmon subwavelength optics. *Nature* 424, 824–830.

Bhatia, P., B.D. Gupta. 2013. Surface plasmon resonance based fiber optic ammonia sensor utilizing bromocresol purple. *Plasmonics* 8, 779–784.

Bilro, L., N. Alberto, J.L. Pinto, R. Nogueira. 2012. Optical sensors based on plastic fibers. *Sensors* 12, 12184–12207.

Chou, S.Y., P.R. Krauss, P.J. Renstrom. 1995. Imprint of sub–25 nm vias and trenches in polymers. *Appl. Phys. Lett.* 67, 3114–3116.

Consales, M., A. Ricciardi, A. Crescitelli, E. Esposito, A.C. Cutolo, A. Cusano. 2012. Lab-on-fiber technology: Toward multifunctional optical nanoprobes. *ACS Nano.* 6, 3163–3170.

Culshaw, B. 2004. Optical fiber sensor technologies: opportunities and—perhaps-pitfalls. *J. Lightwave Technol.* 22, 39–50.

Cusano, A., J.M. López-Higuera, I.R. Matias, B. Culshaw. 2008. Editorial optical fiber sensor technology and applications. *IEEE Sens. J.* 8, 1052–1054.

Decher, G. 1997. Fuzzy nanoassemblies: Toward layered polymeric multicomposites. *Science* 277, 1232–1237.

Downes, F., C.M. Taylor. 2017. Theoretical investigation into the optimization of an optical fiber surface plasmon resonance hydrogen sensor based on a PdY alloy. *Meas. Sci. Technol.* 28, 01510411p.

Elosua, C., C. Bariain, I.R. Matias. 2012. Optical fiber sensing applications: detection and identification of gases and volatile organic compounds. pp. 27–52. *In*: Yasin, M., W.H. Sulaiman, A. Hamzah (eds.). Fiber Optic Sensors. Intech.

Elosua, C., D. Lopez-Torres, M. Hernaez, I.R. Matias, F.J. Arregui. 2013. Comparative study of layer-by-layer deposition techniques for poly(sodium phosphate) and poly(allylamine hydrochloride). *Nanoscale Res. Lett.* 8, 539.

Elosua, C., F.J. Arregui, I.D. Villar, C.R. Zamarreño, J.M. Corres, C. Bariain, J. Goicoechea, M. Hernaez, P.J. Rivero, A.B. Socorro, A. Urrutia, P. Sanchez, P. Zubiate, D.L. Torres, N.D. Acha, J. Ascorbe, A. Ozcariz, I.R. Matias. 2017. Micro and nanostructured materials for the development of optical fibre sensors. *Sensors* 17, 2312.

Fu, H., Y. Jiang, J. Ding, J. Zhang, M. Zhang, Y. Zhu, H. Li. 2018. Zinc oxide nanoparticle incorporated graphene oxide as sensing coating for interferometric optical microfiber for ammonia gas detection. *Sens. Actuators B* 254, 239–247.

Genov, D.A., A.K. Sarychev, V.M. Shalaev, A. Wei. 2004. Resonant field enhancements from metal nanoparticle arrays. *Nano Lett.* 4, 153–158.

González-Sierra, N., L. Gómez-Pavón, G. Pérez-Sánchez, A. Luis-Ramos, P. Zaca-Morán, J. Muñoz-Pacheco, F. Chávez-Ramírez. 2017. Tapered optical fiber functionalized with palladium nanoparticles by drop casting and laser radiation for H_2 and volatile organic compounds sensing purposes. *Sensors* 17, 2039.

Hosoki, A., M. Nishiyama, H. Igawa, A. Seki, Y. Choi, K. Watanabe. 2013. A surface plasmon resonance hydrogen sensor using $Au/Ta_2O_5/Pd$ multi-layers on hetero-core optical fiber structures. *Sens. Actuators B* 185, 53–58.

Iler, R.K. 1966. Multilayers of colloidal particles. *J. Colloid Interface Sci.* 21, 569–594.

James, S.W., R.P. Tatam. 2006. Fibre optic sensors with nano-structured coatings. *J. Opt. A Pure. Appl. Opt.* 8, 430–444.

Jaroszewicz, L.R., B. Culshaw, A.G. Mignani. 2005. Proceedings SPIE 5952. Optical Fibers: Applications, 595201.

Jung, I.W., B. Park, J. Provine, R.T. Howe, O. Solgaard. 2011. Highly sensitive monolithic silicon photonic crystal fiber tip sensor for simultaneous measurement of refractive index and temperature. *J. Lightw. Technol.* 29, 1367–1374.

Kim, J., K. Yong. 2011. Mechanism study of ZnO nano rod-bundle sensors for H_2S gas sensing. *J. Phys. Chem. C* 115, 7218–7224.

Kim, W.S., B.S. Lee, D.H. Kim, H.C. Kim, W.R. Yu, S.H. Hong. 2010. SnO_2 nanotubes fabricated using electrospinning and atomic layer deposition and their gas sensing performance. *Nanotechnology* 21, 1–7.

Koike, Y., K. Koike. 2011. Progress in low-loss and high-bandwidth plastic optical fibers. *J. Polym. Sci. Part B Polym. Phys.* 49, 2–17.

Korposh, S., S. Kodaira, R. Selyanchyn, F.H. Ledezma, S.W. James, S. Lee. 2018. Porphyrin-nano assembled fiber-optic gas sensor fabrication: optimization of parameters for sensitive ammonia gas detection. *Opt. Laser Technol.* 101, 1–10.

Lee, B. 2003. Review of the present status of optical fiber sensors. *Opt. Fiber Technol.* 9, 57–79.

Li, X., W. Cai, J. An, S. Kim, J. Nah, D. Yang, R. Piner, A. Velamakanni, I. Jung, E. Tutuc, S.K. Banerjee, L. Colombo, R.S. Ruoff. 2004. Large-area synthesis of high-quality and uniform graphene films on copper foils. *Science* 306, 1312–1314.

Liehr, S. 2011. Polymer Optical Fiber Sensors in Structural Health Monitoring; Springer: New York, NY, USA, Volume 96, ISBN 9783642210983.

Lin, Y., Y. Zou, R.G. Lindquist. 2011. A reflection-based localized surface plasmon resonance fiber-optic probe for biochemical sensing. *Biomed. Opt. Exp.* 2, 478–484.

Liu, J., Y. Sun, X. Fan. 2009. Highly versatile fiber-based optical Fabry–Pérot gas sensor. *Opt. Exp.* 17, 2731–2738.

Liu, Y., Y.P. Chen, H. Song, G. Zhang. 2012. Modeling analysis and experimental study on the optical fiber hydrogen sensor based on Pd–Y alloy thin film. *Rev. Sci. Instrum.* 83, 1–5.

Llobet, E. 2013. Gas sensors using carbon nanomaterials: a review. *Sens. Actuators B* 179, 32–45.

Manivannan, S., A.M. Saranya, B. Renganathan, D. Sastikumar, G. Gobi, K.C. Park. 2012. Single-walled carbon nanotubes wrapped poly–methyl methacrylate fiber optic sensor for ammonia, ethanol and methanol vapors at room temperature. *Sens. Actuators B* 171-172, 634–638.

Mishra, S.K., B.D. Gupta. 2012. Surface plasmon resonance–based fiber–optic hydrogen gas sensor utilizing indium–tin oxide (ITO) thin films. *Plasmonics* 7, 627–632.

Mishra, S.K., S.N. Tripathi, V. Choudhary, B.D. Gupta. 2014. SPR based fiber optic ammonia gas sensor utilizing nanocomposite film of PMMA/reduced graphene oxide prepared by *in situ* polymerization. *Sens. Actuators B* 199, 190–200.

Mishra, S.K., B.D. Gupta. 2015. Surface plasmon resonance–based fiber optic chlorine gas sensor utilizing indium–oxide–doped tin oxide film. *IEEE J. Lightwave Technol.* 33, 2770–2776.

Mishra, S.K., S. Bhardwaj, B.D. Gupta. 2015. Surface plasmon resonance–based fiber optic sensor for the detection of low concentrations of ammonia gas. *IEEE Sens. J.* 15, 1235–1239.

Mishra, S.K., S.N. Tripathi, V. Choudhary, B.D. Gupta. 2015. Surface plasmon resonance-based fiber optic methane gas sensor utilizing graphene-carbon nanotubes-poly(methyl methacrylate) hybrid nanocomposite. *Plasmonics* 10, 1147–1157.

Moon, H.G., Y.S. Shim, D. Su, H.H. Park, S.J. Yoon, H.W. Jang. 2011. Embossed TiO_2 thin films with tailored links between hollow hemispheres: synthesis and gas sensing properties. *J. Phys. Chem. C* 115, 9993–9999.

Nanda, S.S., D.K. Yi, K. Kim. 2016. Study of antibacterial mechanism of graphene oxide using Raman spectroscopy. *Sci. Rep.* 6, 28443.

Pandey, P.A., N.R. Wilson, J.A. Covington. 2013. Pd-doped reduced graphene oxide sensing films for H_2 detection. *Sens. Actuators B, Chem.* 183, 478–487.

Pathak, A., S.K. Mishra, B.D. Gupta. 2015. Fiber-optic ammonia sensor using Ag/SnO_2 thin films: optimization of thickness of SnO_2 film using electric field distribution and reaction factor. *Appl. Opt.* 54, 8712–8721.

Raj, D.R., S. Prasanth, T.V. Vineeshkumar, C. Sudarsanakumar. 2015. Ammonia sensing properties of tapered plastic optical fiber coated with silver nanoparticles/PVP/PVA hybrid. *Opt. Comm.* 340, 86–92.

Renganathan, B., D. Sastikumar, G. Gobi, N.R. Yogamalar, A.C. Bose. 2011. Nanocrystalline ZnO coated fiber optic sensor for ammonia gas detection. *Opt. Laser Technol.* 43, 1398–1404.

Robinson, J.T., F.K. Perkins, E.S. Snow, Z. Wei, P.E. Sheehan. 2008. Reduced graphene oxide molecular sensors. *Nano Lett.* 8, 3137–3140.

Rosli, M.A.A., P.T. Arasu, A.S.M. Noor, H.N. Lim, N.M. Huang. 2016. Reduced graphene oxide nano-composites layer on fiber optic tip sensor reflectance response for sensing of aqueous ethanol. *J. Eur. Opt. Soc.* 12, 22.

Sanders, M., Y. Lin, J. Wei, T. Bono, R.G. Lindquist. 2014. An enhanced LSPR fiber-optic nanoprobe for ultrasensitive detection of protein biomarkers. *Biosensors Bioelectron.* 61, 95–101.

Scheerlinck, S., P. Bienstman, E. Schacht, D.V. Thourhout. 2009. Metal grating patterning on fiber facets by UV-based nano imprint and transfer lithography using optical alignment. *J. Lightw. Technol.* 27, 1415–1420.

Shobin, L.R., D. Sastikumar, S. Manivannan. 2014. Glycerol mediated synthesis of silver nanowires for room temperature ammonia vapor sensing. *Sens. Actuators A* 214, 74–80.

Son, D., S.J. Kim, S. Lee. 2017. Electrospun assembly: A nondestructive nanofabrication for transparent photosensors. *Nanotechnology* 28, 155202.

Strobel, O., D. Seibl, J. Lubkoll, R. Rejeb. 2009. ICTON 2009. 11th International Conference on Transparent Optical Networks.

Sun, K., N. Wu, C. Guthy, X. Wang. 2011. Nanomaterial fiber optic sensors in healthcare and industry applications. pp. 163–170. *In*: Lu, K., N. Manjooran, M. Radovic (eds.). Advances in Nanomaterials and Nanostructures. New Jersey: John Wiley, 229.

Tabassum, R., B.D. Gupta. 2015. Surface plasmon resonance based fiber optic detection of chlorine utilizing polyvinylpyrolidone supported zinc oxide thin films. *Analyst* 140, 1863–1870.

Tabassum, R., B.D. Gupta. 2015. Surface plasmon resonance–based fiber–optic hydrogen gas sensor utilizing palladium supported zinc oxide multilayers and their nanocomposite. *Appl. Opt.* 54, 1032–1040.

Tiwari, D., K. Mullaney, S. Korposh, S.W. James, S.W. Lee, R.P. Tatam. 2017. An ammonia sensor based on lossy mode resonances on a tapered optical fiber coated with porphyrin–incorporated titanium dioxide. *Sens. Actuators B* 242, 645–652.

Urrutia, A., J. Goicoechea, P.J. Rivero, I.R. Matías, F.J. Arregui. 2013. Electrospun nanofiber mats for evanescent optical fiber sensors. *Sens. Actuators B Chem.* 176, 569–576.

Usha, S.P., S.K. Mishra, B.D. Gupta. 2015. Fabrication and characterization of a SPR based fiber optic sensor for the detection of chlorine gas using silver and zinc oxide. *Materials* 8, 2204–2216.

Valerini, D., A. Cretì, A.P. Caricato, M. Lomascolo, R. Rella, M. Martino. 2010. Optical gas sensing through nanostructured ZnO films with different morphologies. *Sens. Actuators B* 145, 167–173.

Viheriala, J., T. Niemi, J. Kontio, T. Rytkonen, M. Pessa. 2007. Fabrication of surface reliefs on facets of single mode optical fibres using nanoimprint lithography. *Electron. Lett.* 43, 150–151.

Vikrant, K., V. Kumar, K.-H. Kim. 2018. Graphene materials as a superior platform for advanced sensing strategies against gaseous ammonia. *J. Mater. Chem. A* 6, 22391–22410.

Wang, X., C. Drew, S.H., Lee, K.J. Senecal, J. Kumar, L.A. Samuelson. 2002. Electrospun nanofibrous membranes for highly sensitive optical sensors. *Nano Lett.* 2, 1273–1275.

Wang, X., Y. Tang, C. Zhou, B. Liao. 2013. Design and optimization of the optical fiber surface plasmon resonance hydrogen sensor based on wavelength modulation. *Opt. Comm.* 298-299, 88–94.

Wang, Z.L. 2004. Zinc oxide nanostructures: growth, properties and applications. *J. Phys.: Condens. Matter* 16, 829–858.

Xiang, Q., G.F. Meng, H.B. Zhao, Y. Zhang, H. Li, W.J. Ma, J.Q. Xu. 2010. Au nanoparticle modified WO_3 nanorods with their enhanced properties for photocatalysis and gas sensing. *J. Phys. Chem. C* 114, 2049–2055.

Xu, J., J. Han, Y. Zhang, Y. Sun, B. Xie. 2008. Studies on alcohol sensing mechanism of ZnO based gas sensors. *Sens. Actuators B* 132, 334–339.

Yamini, K., B. Renganathan, A.R. Ganesan, T. Prakash. 2017. Clad modified optical fiber gas sensors based on nanocrystalline nickel oxide embedded coatings. *Opt. Fiber Technol.* 36, 139–143.

Yao, B., Y. Wu, Y. Cheng, A. Zhang, Y. Gong, Y.-J. Rao, Z. Wang, Y. Chen. 2014. All-optical Mach-Zehnder interferometric NH_3 gas sensor based on graphene/microfiber hybrid waveguide. *Sens. Actuators B* 194, 142–148.

Yu, C., Y. Wu, X. Liu, F. Fu, Y. Gong, Y.J. Rao, Y. Chen. 2017. Miniature fiber-optic NH_3 gas sensor based on Pt nanoparticle-incorporated graphene oxide. *Sens. Actuators B* 244, 107–113.

Zhang, J., S. Wang, M. Xu, Y. Wang, B. Zhu, S. Zhang, W. Huang, S. Wu. 2009. Hierarchically porous ZnO architectures for gas sensor application. *Cryst. Growth Des.* 9, 3532–3537.

Chapter **5**

Carbon Nanomaterials as Chemical Sensors

1. Introduction

In the last few years, there has been an explosion in the number of researchers dealing with nano-materials in chemical sensing. However, there is a requirement for a simple, sensitive and stable electronic sensing probe suited for trace detection in a wide spectrum of applications that ranges from lab-on-a-chip and *in vivo* bio-sensors in environmental monitoring and warfare agent detection, as opposed to the expensive, bulky and complicated instrumental technologies. Therefore, more focused research work should be taken into account to develop inexpensive, low-power devices. Carbon nanomaterials have been extensively utilized as sensing materials for their excellent detection sensitivity with interesting transduction characteristics in a single-layer material (Roberts et al. 2009). It is well known that low dimensional carbon structures have most of their exposure to the environment, which will result in high specific surface area and greater sensitivity. Few nanomaterials such as carbon nanotubes or graphene have high-quality crystal lattices and show high carrier mobility and lower noise. Single crystalline nano-structures are good model materials for running computational chemistry studies. The sensitivity and selectivity of carbon nanostructured materials can be fabricated by employing various methodologies both to create defects and graft functional groups to their surface in a controlled way. Moreover, their mechanical properties make them potential candidates in becoming integrated with flexible electronic devices. They also can offer high sensitivity to cost ratio, operated at room temperature. Their low power consumption makes them a suitable candidate for being operated remotely. The electronic signal transduction of chemical environmental analytes is advantageous over optical technologies due to their economic nature, simple device, greater sample throughput and better portability. In the last few years, carbon nanotubes became the most studied carbon nano-materials as effective sensors. Recently, graphene has taken the position of nanotubes. Carbon nanomaterials exist as nanoparticles, nanospheres, fibers, wires, etc. This chapter will first discuss the extensive synthetic strategies of nanospheres and nanotubes. This will also review the role of carbon nano-materials as the new generation chemical sensing probes with superior performance and discuss their advantages and challenges.

2. Synthetic Strategies of Carbon Nano-spheres

In recent days, functional carbon nanospheres with controllable size, special surface morphology, porosity and chemical composition are the effective reasons to make them potential candidate in a wide variety of applications, viz., catalysts or catalytic support, electrode materials and adsorbents (Roberts et al. 2014, Liang et al. 2012). Porous carbon nanospheres can provide higher pore accessibility and faster molecular diffusion-transfer, and they have critical importance in carbon-based methods. Moreover, carbon supports provide extraordinary high stability in both acidic and basic environments, and the easy recovering of noble metals by calcination of the carbon after catalyst deactivation. Ultra-file carbon nanospheres with a size below 200 nm could be easily internalized into cells by intracellular endocytosis and therefore have been successfully extended to biomedical and pharmaceutical applications in respect to imaging agents, delivery of drugs, genes and proteins (Gu et al. 2011). In recent times, Wang et al. used hollow carbon nanospheres as a novel platform in delivering the drug doxorubicine and generation of additional cellular reactive oxygen species under near IR laser irradiation. Further, these irradiated carbon nanospheres produced a large amount of heat shock factor – 1 protein homotrimers via catalyzed persistent free radicals (Wang et al. 2015). Carbon-spheres have been reported to synthesize using chemical vapor deposition (CVD), nano-casting with silica spheres via hard templates, hydrothermal carbonization of sugars, modified Stober method, soft-templating methodologies via organic-organic self-assembly and so on by many research groups. Usually, CVD has been used in the presence of metal oxides as catalysts, therefore purification is needed to mitigate the presence of a catalyst. This results in the limitation of this method to apply on small scale. On the other hand, nano-casting methodology mostly has been applied with silica as a template. Generally, carbon nano-spheres have been synthesized using infiltration of the micro-/mesopores of a silica template with a carbon precursor, which have been followed by carbonization at high temperature to form a silica-carbon composite material. In the next step, the silica template has been dissolved and removed under acidic or alkaline conditions. A research group reported CMK-1 type mesoporous carbon nanospheres with an average size of around 150 nm with MCM-48 type mesoporous silica nanoparticles as hard template and furfuryl alcohol as the carbon source (Kim et al. 2008).

(a) Hydrothermal Carbonization of Sugars

Hydrothermal carbonization (HTC) is one of the promising synthetic strategies which was established around a century ago, the time when Bergius experimented with the hydrothermal transformation of cellulose into coal-like materials (Bergius et al. 1913). In this method, artificial acceleration of carbohydrate decomposition, polycondensation and polymerization in the case of highly cross-linked and carbon-rich polymers at moderate temperature (< 300°C) and self-generated pressure (~ 1.0 MPa). Mostly in this method, glucose, fructose, cellulose and raw biomass have been utilized in the feasible transformation process (Titirici et al. 2012). HTC of these carbohydrates successfully decomposed into hydroxymethyl-furfural and further processed with polymerization-polycondensation via the formation of small nuclei. These nuclei have been reported to grow through the LaMer model until all the monomer molecules have been completely consumed, and the final particle size has been achieved properly. On the other hand, direct HTC of carbohydrates precursors with a concentration of 10–30 wt% with a diameter of above 500 nm. However, these carbon nanospheres usually formed without significant porous character, therefore various additives or catalysts have been reported to introduce into the HTC process to change the porous character (White et al. 2014, Liang et al. 2012, Sevilla et al. 2011, Yu et al. 2012).

• Bulk Composition

Initially, the methodology initiated with the number of biomass-derived precursors including simple molecules (viz., hydroxymethyl furfural), sugars (viz., glucose, fructose, xylose, cellulose, starch) and raw materials (viz., microalgae, pine needle) (Titirici et al. 2007, Falco et al. 2012, Mane et al. 2012, Sevilla et al. 2009, Sevilla et al. 2012, Unur et al. 2013, Roman et al. 2013, Titrici et al. 2007, Xie et al. 2011). In this typical synthesis method, the carbon sphere size varies from 500 nm to 10 μm. Additionally, several factors are there to control the size of the spheres, such as the concentration of precursors and reaction temperature. Titirici et al. had shown that with an increase in the reaction temperature, the size of the spheres got increased. They had compared the synthesis at two different temperatures of 260°C and 160°C in which the size had been revealed as 685 nm and 474 nm, respectively. Moreover, higher temperature led to the formation of spheres with a more homogeneous size distribution. The principle behind this phenomena had been explained as a substantial change in solvent ability and reacting behavior through more heat treatment in spheres, which would undergo a deeper polymerization under an enhanced amount of self-generated pressure. HTC is also a wide technique to produce functionalized carbon nanospheres, as waster soluble vinyl monomer molecules with specific function could be added to the HTC mixture via coupling process with hydroxymethylfurfural. This total process has resulted in the formation of carbon nanospheres consisting of functional groups on the surface. In the case of glucose HTC in presence of acrylic acid afforded with carboxylate group rich carbonaceous nanospheres. These carbon nanospheres have been successfully implied in heavy metal removals from aqueous solution (Demir-Cakan et al. 2009). Besides, the addition of acrylic acid has induced a change in the particle morphology via the formation of smaller aggregated particles of 250 nm size. Additionally, these acrylic acid molecules have helped to stabilize the initial small droplets. Many works have been reported in which imidazole group has been added during HTC of glucose, which resulted in the transformation of imidazole moieties on the surface into the corresponding butyl imidazoliums with bromine ions like ionic liquids attached to carbon nanospheres (Demir-Cakan et al. 2010). These types of carbon nanospheres have been extensively used as catalysts in organic transformations (transesterification) and Diels-Alder or Knoevenagel condensation reactions.

• Porous Composition

In this methodology, the direct HTC technique is executed, therefore inherently solid carbon nanospheres are formed. The porosity of the spheres is limited from the outer surface. In recent days, a numerous number of synthetic strategies have been adopted in which the suitable additives developed which controlled the sphere size with significant porous structure (Hu et al. 2010, Falco et al. 2013, Zhao et al. 2010, Zhang et al. 2014). This strategy particularly provides easy way to the direct synthesis of porous hydrothermal carbon-based nanospheres.

One research group reported HTC of glucose with ovalbumin additives, resulting in sponge-like nanostructured carbon spheres with a size range of 20 to 50 nm (Baccile et al. 2010). Ovalbumin is a simple and accessible protein that stabilized the surfaces of smaller carbon nanoparticles, which could be unstable and aggregate toward linked networks with mesoporous domains. BET surface area value of HTC carbon has been reported to reach 110 $m^2.g^{-1}$ (White et al. 2011). Another research group developed an inexpensive, abundant sodium borate (borax) as a novel complex structure-directing agent for nanostructured carbon monoliths made up of aggregated carbon nanospheres (Fellinger et al. 2012). In this case, by using relatively high borax concentration, the diameter of the nanospheres could be reduced to < 7 nm. Those aggregated nanospheres could form the desired hierarchical pore system, which could be potentially applied as heterogeneous catalysis and electrochemical reactions. This type of process reported BET surface area of up to 233 $m^2.g^{-1}$ for HTC carbon at 180°C and 614 $m^2.g^{-1}$ with treatment at 550°C. Few researchers utilized nitrogen-rich ionic polymer as an additive to develop a series of N-doped carbon nanospheres in support of Pd nanoparticles. These Pd-N-carbon nanospheres afforded an extraordinary performance in the solvent-free oxidation process of hydrocarbons and alcohol with air as an oxidant (Zhang et al. 2013). In this typical method, the doped N atoms with lone pair of electrons present on the surface might have enhanced the e^- density of Pd^0 and further promoted the oxidative addition reaction of the O-H bond or the C-H bond to the Pd^0 species, led to the enhancement of reaction rate. There have been few reports who have introduced the polymerized ionic liquids (PILs) as smart additives during the HTC process to stable the primary structure of carbon nanospheres even with the application of very small amounts, i.e., 0.5 wt% (Zhang et al. 2013). In this method, the polymer chains stabilized the primary nanospheres produced at the initial stage and allowed the growth further with the additions of monomer molecules. Additionally, the charge feature of the PILs initiated the electrostatic repulsion to stabilize the structure of nanospheres and diminished their agglomeration. These fructose-based nanospheres under HTC at 160°C could attain BET specific surface areas of up to 161 m^2 g^{-1} with mesoporous structures to some extent. Particularly, this methodology could successfully and potentially apply in the formation of Au-Pd core-shell structured nanoparticles, which were supported on carbon nanospheres via a one-pot process. Apart from this, a few homo-disperse latex nanoparticles were also interestingly utilized in the synthesis of carbon hollo-spherical materials. One research group reported a facile, inexpensive and sustainable synthetic strategy in which glucose-derived hollow carbonaceous nanospheres were considered with the application of hydroxyl-terminated polystyrene latex nanoparticles. They were selected as suitable hydrogen bonding surfaces on which the decomposition and polymerization of glucose molecules could proceed (White et al. 2010).

The materials further were heated to the desired temperature range to remove the polymer template material with simultaneous graphitization of the carbonaceous shell structure. Following this method, uniform hollo carbon nanospheres could be synthesized with an internal diameter of ~ 80 nm with a shell thickness of ~ 10 nm. Moreover, these hollo spheres possessed a high surface area value of 350 $m^2.g^{-1}$ with pore volume above 0.6 $cm^3.g^{-1}$. In this case, these carbonaceous hollow-spheres had been utilized in a sodium-based battery and exhibited an excellent cycling activity and rate capability (Tang et al. 2012). However, these polymers are usually unstable at the typical HTC process temperature, i.e., 180–240°C, one research group had introduced fructose as a carbon source for micellar self-assembly at 130°C, further the Pluronic F127 was removed with thermal treatment at 550°C in a N_2 atmosphere (Kubo et al. 2011). In this work, BET specific surface area was reported as 257 $m^2. g^{-1}$. The pore size distribution results had shown a sharp peak at 0.9 nm for microporous materials.

Thermal treatment is considered an important key factor in the HTC process. Usually, 550–1,100°C temperature under nitrogen or argon atmosphere is applied during thermal treatment, which can produce microporous carbon materials. Few research groups had adopted different temperature ranges (at 900°C in static air atmosphere) to develop hierarchical carbon materials. In this typical process, the acceptable carbon yield reached much higher (~ 23%) in comparison to that obtained from KOH activation. A slight aggregation could be possible to happen with BET-specific surface area value up to 1,306 $m^2.g^{-1}$ (Gong et al. 2014). In this particular method, the heating process took place an important role where the HTC of sugar processed at a relatively lower range of temperature, e.g., 130–200°C, followed by thermal carbonization at a higher temperature of 600–1,000°C with higher carbon content. Here, high-temperature application of sucrose resulted in many interesting outcomes. In a typical method, direct pathway to hollow carbon nanospheres formation following pressure-assisted carbonization at 550°C of sucrose in a closed autoclave with Zn as a reductant in which the diameter around 50–100 nm with 270 $m^2.g^{-1}$ of surface area (Han et al. 2011). The duration of thermal treatment also plays a pivotal role in the resultant carbon spheres. In the conventional HTC method, thermal treatment usually processed using an autoclave for 8 to 24 hours duration. On the other hand, microwave-assisted heat treatment could proceed with uniform heat resulting in the achievement of the desired temperature in lesser time. The technology proceeded with selective transferring energy into the microwave absorbing polar solvents. One research group had adopted a synthetic approach in which a simple microwave-assisted reaction of sucrose had been followed by using water-ethylene glycol as the solvents (Chen et al. 2014).

(b) Application of Friedel-Craft Reaction-Induced Polyaromatic Precursors

In 2013, a versatile synthetic approach had been taken into account in which hyper-cross-linked porous polymer had been considered in Friedel-Crafts alkylation of aromatic monomers using a cross-linker in the presence of a Lewis acid (Xu et al. 2013). In this typical process, a template-free approach had been taken into consideration to control their shape into spheres. Ouyang et al. reported highly monodisperse microporous carbon nanospheres with a diameter as low as 190 nm using a facile hyper-cross-linking strategy (Ouyang et al. 2013). In this case, monodisperse polystyrene nanospheres with divinylbenzene pre-crosslinking had been initially utilized as precursor materials, which was followed by a simple hyper-cross-linking method using Friedel-Crafts reaction of polystyrene chains in the presence of carbon tetrachloride (as cross-linker) and $AlCl_3$ (as a catalyst). These carbon nanospheres had been formed with a highly microporous structure with a BET surface area value of 1,357 $m^2.g^{-1}$. In another work, polyaromatic precursors had been produced using aromatic hydrocarbons, viz., pyrene, naphthalene, anthracene and pheanthracene using $ZnCl_2$ as the catalyst and chloromethyl methyl ether as the cross-linker material (Huang et al. 2013). This method had produced a grape-like network with BET surface area value up to 455 $m^2.g^{-1}$.

In summary, we can say that Friedel-Crafts alkylation of aromatic monomers with applying a cross-linker can develop an overall simple strategy to polymer carbon nanospheres with a uniform size distribution. In this strategy, all the precursor materials are abundant and this can be adopted without the requirement of any template. Moreover, the pore size of these nanospheres can easily be controlled, and it can be a great way to synthesize carbon nanospheres with versatile characteristics.

(c) Modified Stöber Method

In 1968, the Stöber method had been established to synthesize silica spheres with controllable diameter and uniform particle size (Stöber et al. 1968). It is a very simple synthetic strategy using ammonia catalyzed hydrogen process and followed by the condensation of tetraethyl orthosilicate (TEOS) in ethanol-water mixtures. This method had been further adopted to synthesize ordered mesoporous carbons in which resorcinol-formaldehyde had been used as promising carbon precursors by following soft-templating technology. Especially, this method had been taken into consideration in which carbon precursors with rich hydrogen bonding were required (Liang et al. 2008). In this typical method, resorcinol and formaldehyde proceed with the following steps: (i) addition reaction to produce hydroxymethyl (–CH2OH) derivatives of resorcinol, (ii) condensation reaction of the hydroxymethyl derivatives to form methylene (–CH2-) and methyelene ether (–CH2OCH2-) bridged compounds via elimination of water, which results into the formation of polymerized resorcinol-formaldehyde resin network structure. In 2011, one research group had initiated the controlled synthetic method of carbon nanospheres via ammonia promoted polymerization reaction of resorcinol and formaldehyde in an ethanol-water solution, followed by the thermal treatment under nitrogen atmosphere (Liu et al. 2011). In this case, the average size of the carbon nanospheres had been revealed as 500–900 nm. Figure 5.1 represents the TEM image of these carbon nanospheres. In this modified Stöber method, the carbonization of primary polymer spheres with cross-linked resorcinol–formaldehyde network might form direct porosity to some degree via decomposition of unstable species.

In another work, the researchers investigated the effect of reaction time, ethanol concentration and carbonization temperature during this synthetic strategy, and they studied the process in ion storage/transfer behavior of carbon nanospheres in an electrical double-layer capacitor (Tanaka et al. 2012).

Several combined routes had been reported in the literature regarding the modified Stöber method (Ma et al. 2014, Choma et al. 2014, Zhao et

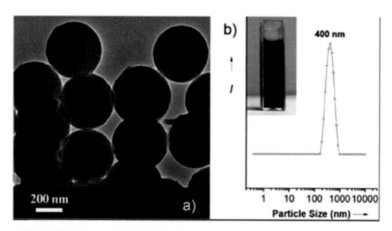

Figure 5.1: (a) TEM images and (b) DLS plot of carbon spheres obtained by carbonizing the 520 nm RF colloidal spheres (inset: photograph illustrating the dispersivity of the carbon spheres in ethanol) (Reproduced with permission from Liu et al. 2011).

al. 2013, You et al. 2012). In a work, Pluronic F127 block copolymer was introduced into this method, which resulted in the smaller sized carbon nanospheres with an average diameter of 30–100 nm (Choma et al. 2012). After synthesis carbon nanospheres had been activated with KOH, $K_2Cr_2O_4$ or CO_2 was also introduced, during which micro-porosity was greatly enlarged with retainment of spherical shape (Ludwinowicz et al. 2015, Zhang et al. 2014, Wickramaratne et al. 2013). In recent years, hetero atom incorporated (N, B, F and S) carbon materials have drawn huge attention. The heteroatoms can transform the carbon properties, and it can afford a carbon material with attractive characteristics for many more uses, such as electrodes for oxygen reduction reactions, metal-free catalysis and CO_2 capture (Wei et al. 2013, Zhang et al. 2013, Zhang et al. 2014).

Another research group had reported nitrogen-incorporated highly porous carbon nanospheres using a modified Stöber synthesis by applying melamine, which acted as a nitrogen precursor material. In this work, nitrogen had been incorporated up to ~ 4% which led to porous solids with high absorption capacity toward CO_2 gas (6.9 mmol.g^{-1} at 0°C) (Wickramaratne et al. 2014). In another work, nitrogen content reached up to 14.8 wt% by using SiO_2 spheres as hard templates and resorcinol-formaldehyde as a carbon precursor via two steps of the Stöber method. In this typical pathway, silica colloidal spheres were produced via the classical Stöber method and further *in situ* acted as core material for the deposition of resorcinol-formaldehyde shells uniformly. With co-carbonization with melamine (used as nitrogen resource) and removal of cores, an N-doped carbon sphere was obtained which successfully had been utilized in CO_2 capture (2.67 mmol g^{-1} at room temperature) (Feng et al. 2014). Another important method had been reported in which polymerization of 3-amino-phenol and formaldehyde took place in a mixture of water and ethanol with cystine as a catalyst (Yang et al. 2014). Other hetero atoms such as boron or sulfur had also been incorporated into the carbon nanospheres by using dibenzyl disulfide or boric acid as an additive. Sometimes, urea had been successfully used as an additive in high nitrogen content carbon nanospheres (Song et al. 2014). Recently, one research group carried out polymerization of dopamine under the modified Stöber method in which uniform nitrogen-containing carbon nanospheres had been produced with diameter from 120 to 780 nm (Ai et al. 2013).

3. Synthetic Strategies of Carbon Nanotubes

Nanotubes belong to a promising group of nanomaterials. Carbon nanotubes contain one or several concentric graphite-like layers with a diameter range of 0.4 nm–10 nm. In 1991, Iijima discovered carbon nanotubes with an experimental observation by TEM and successfully developed large quantities of nanotubes (Ijima et al. 1991). Carbon nanotubes (CNTs) can be described as graphite sheets that rolled up into a cylindrical shape. CNTs can be produced by three different methods arc discharge method, laser ablation method and chemical vapor deposition method. They possess excellent mechanical, electronic, thermal, optical and chemical properties which have successfully revolutionized the state-of-the-art in nanotechnology. CNTs have been broadly classified into three groups: Single-Walled Nanotubes (SWNTs), Double-Walled Nanotubes (DWNTs) and Multi-walled Nanotubes (MWNTs).

(a) Arc Discharge Method

Arc discharge can be defined as the electrical breakdown of gas to generate plasma.

This old technique of generating arc using electric current was primarily utilized by Ijima to develop CNTs first time in 1991 (Ijima et al. 1991).

A schematic diagram of an arc discharge chamber is represented in Figure 5.2 in which the chamber is consisting of 2 electrodes, which are placed either horizontally or vertically. One of which (anode) is usually filled with powdered carbon precursor materials (graphite) along with the suitable catalyst and the other with an electrode, i.e., cathode is usually a pure graphite rod. The chamber is usually filled up with gas or submerged

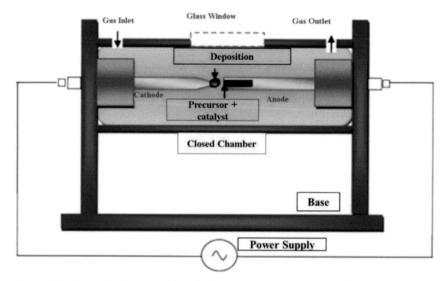

Figure 5.2: Schematic diagram of Arc discharge set-up (reproduced with permission from Arora et al. 2014).

inside a liquid environment (water, ethanol, etc.). After switching on the power supply (AC or DC), the electrodes are brought into contact with each other to generate an arc and are placed at an intermittent gap of 1–2 nm to attain a steady discharge. A constant is used to maintain through the electrodes to get a non-fluctuating arc for which closed-loop automation is generated to adjust the gap automatically. Generally, a fluctuating arc can result in unstable plasma and therefore the quality of the resultant product gets affected. The arc current produces plasma of very high temperature (~ 4,000–6,000 K), which can sublime the carbon precursor at the anode, which in turn produce carbon vapors. These vapors aggregate into the gas phase and drift toward the cathode. At the cathode, it gets cooled down due to the temperature gradient. After an arc application time of a few minutes, the discharge is stopped and the cathodic deposit which contains CNTs along with the soot is collected from the walls of the chamber. The deposit is further purified and analyzed under an electron microscope to investigate its morphological structure (Arora et al. 2014). In literature, many research groups have been described the mechanism of nanotube synthesis (Liang et al. 2012, Wu et al. 2012, Keidar et al. 2007, Farhat et al. 2001, Heer et al. 2005, Gamaly et al. 1995, Ugarte et al. 1994, Zhou et al. 2003). However, the exact growth mechanism is still debatable as different theories are present regarding nanotube development in the vapor phase (Gamely et al. 1994), liquid phase (Heer et al. 2005), a solid phase (Ugarte et al. 1994) and a crystal phase (Zhou et al. 2003).

(b) Pulsed Laser Ablation Method

Pulsed laser ablation in liquid (PLAL) is one of the most attractive and promising methods, which offers plenty of advantages over the other synthetic strategies for its simple, inexpensive, non-catalytic approach, no vacuum required, high purity product, the controllable method in particle size and morphology of the resultant products (Ismail et al. 2017, Ismail et al. 2011). Although this method also exhibited some disadvantages, viz., small production of ablated nanoparticles, large particle aggregations and agglomeration. There are a few key parameters that can influence this synthetic method, such as laser fluence, pulse width, repetition rate and wavelength (Bogaerts et al. 2005). The nanoparticles synthesized by the PLAL method have been utilized in many applications, such as gas photocatalyst, antibacterial and optoelectronic devices. MWCNTs were reported to synthesize in which high purity graphite pellet was placed in the bottom of a glass cell filled with double distilled water (DDW) with a few mm gap above the graphite target surface. In this method, the laser energy was corrected after considering the attenuation of water at two laser wavelengths and laser spot size on the graphite was analyzed. In this case, when a laser beam hits onto the metal target, plasma, vapor and metal micro- or nano-sized droplets can be produced as possible initial products, which further can be reacted with the liquid medium to produce nanoparticles.

The laser ablation process usually considers the vaporization of material from a solid target. In this case, intense localized heating of the surface by laser beam happens, which results in high-velocity ejection and evaporation of the impacted material. The rate and extent of the ablation process depend on the nature and condition of the target material as well as the laser parameters (e.g., wavelength, intensity and pulse duration). In this typical process, laser energy is used for heating the target and vaporizing of the material occurs only beyond this threshold. This threshold is directly controlled by the laser-target interaction. Moreover, laser energy becomes too high. Therefore, it will lead to the case of "spalling" in which chunks of the material are removed rather than a small outer surface (Rinzler et al. 1998).

(c) Chemical Vapor Deposition Method

Chemical Vapour Deposition (CVD) draws huge attention in producing CNTs in recent days. In this typical method, the thermal decomposition of a hydrocarbon vapor is achieved in the presence of a metal catalyst. It can be categorized into thermal CVD and catalytic CVD to distinguish it from many other types of CVD used for various purposes. In 1890, French Scientists observed the formation of carbon filaments during an experiment involving the passage of cyanogens over red-hot porcelain (Schultzenberger et al. 1890). Figure 5.3 represents a schematic diagram of the experimental setup utilized for CNT growth by the CVD process.

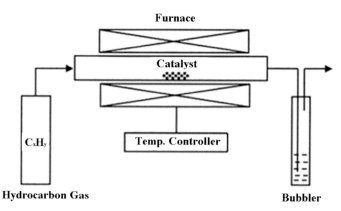

Figure 5.3: Schematic diagram of the experimental set up for CNT by CVD method (reproduced with permission from Kumar et al. 2010).

This process considers bypassing a hydrocarbon vapor for 15–60 min duration via a tubular reactor in which a catalyst present at high temperature (600–1,200°C) to decompose the hydrocarbon. In this method, CNTs grow on the catalyst in the reactor, which is controlled upon cooling down of the system to room temperature. In the case of a liquid hydrocarbon (benzene, alcohol, etc.), the liquid is heated in a flask and an inert gas is purged into it. This system carries the hydrocarbon vapor into the reaction zone. In this case, if a solid hydrocarbon is to be used as the CNT precursor, it can directly be kept in the low-temperature zone of the reaction tube. Volatile materials (camphor, naphthalene, ferrocence, etc.) directly turn from solid to vapor and perform CVD while passing over the catalyst kept in the high-temperature zone. Like the CNT precursors, also the catalyst precursors in CVD may be used in any form: solid, liquid or gas, which may be suitably placed inside the reactor or fed from outside (Kumar et al. 2010).

4. Carbon Nanotubes as Gas Sensors

The electronic properties of CNTs are found to be extremely sensitive to their local chemical environment. This chemical sensitivity has made them potential candidates for incorporation into the development and designing of chemical sensors. Kong et al. had employed SWCNTs as semiconductors in which the CNTs had been synthesized using the CVD method from patterned catalyst islands on SiO_2/Si substrates (Kong et al. 2010, Collins et al. 2010). In this process, two metal contacts were utilized to connect with SWCNT, resulting in the formation of the metal/SWNT/metal system having p-type transistor characteristics with several orders of magnitude change in conductance under different gate voltages (Tan et al. 1998). In this work, sensor conductance was investigated in the presence of NO_2 and NH_3 vapors. The SWNT is a hole-doped semiconductor, where the conductance of the SWNT tends to decrease by three orders of magnitude under positive gate voltages (Collins et al. 2010). Figure 5.4 presents the electrical response of a S-SWNT device to gas molecules. Exposure to NH_3 effectively shifts the valence band of the nanotube away from the Fermi level, leading to hole depletion and reduced conductance. In the case of NO_2 gas, exposure of the initially depleted sample to this gas resulted in the nanotube Fermi level shifting closer to the valence band. This caused enriched hole carriers in the nanotube and enhanced sample conductance. The effects of NO_2 and NH_3 on the electrical properties of mats of SWNT ropes made from as-grown laser ablation materials was also investigated. Response to gases was significantly lower for SWNT than that for single S-SWNT devices. This result was assigned to two reasons: (i) the molecular interaction effects are averaged over

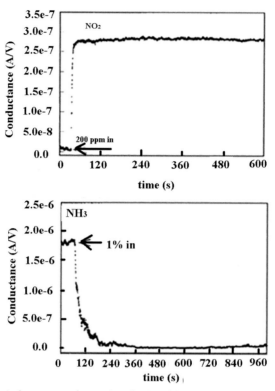

Figure 5.4: Electrical response of a semiconducting SWNT to gas molecules. Conductance (under Vg = +4 V) versus time in a 200-ppm NO₂ flow (upper subplot). Conductance (under Vg = 0 V) versus time recorded with the same S-SWNT sample in a flow of Ar containing 1% NH₃ (lower subplot) (reproduced with permission from Collins et al. 2010).

metallic (i.e., less responsive) and semiconducting tubes and also (ii) the inner tubes in SWNT ropes are blocked from interacting with gaseous materials because molecules are not expected to intercalate into SWNT ropes.

In the same year, Collins et al. had worked on the extreme oxygen sensitivity of SWNT (Collins et al. 2010). Their electrical resistance could be reversibly tuned by exposure to small concentrations of adsorbed O_2 and an apparently semiconducting nanotube could be transformed into an apparent metal via such exposure. They had demonstrated that CNTs could effectively be used as sensitive chemical gas sensors but likewise indicated that many supposedly intrinsic properties measured on as-prepared CNTs might be severely compromised by extrinsic air exposure effects. They had also concluded that any oxygen-induced charge transfer in SWNTs indicated the presence of on-tube defects and the possibility of O_2 absorbance on the inside of CNTs with open ends should also be considered.

Many research groups had already reported functionalization of carbon nanotube sidewalls as the formation of a strong chemical bond between a specific chemical species, and the nanotube can effectively be applied and the selectivity of the adsorption process can also be enhanced in this way (Kauffman et al. 2008, Bondavalli et al. 2009). In a work, Pd-coated CNT had been reported as selectively sensitive toward hydrogen (Kong et al. 2001). The concept of CNT-metal clusters hybrids. A theoretical study had also been reported with considering the CNT-Al cluster (Zhao et al. 2005). It was shown that the adsorption of NH_3 generated a substantial polarisation and charge accumulation in the region between the Al cluster and the nanotube. This provided important information regarding the system's electronic response. However, it affected the ionic component of the bonding with alteration of the position at the Fermi level and the band alignment. Thus, the change in the electrical conductance of the CNT-Al system is a measure of the sensitivity of chemical sensors based on this material (Zhao et al. 2005). Star and co-workers developed CNTs by selective electroplating with different metals, viz., Pt, Pd, Au, Rh, Sn, Mg, Fe, Co, Ni, Zn, Mo, W, V and Cr for detecting CO, NO_2, CH_4, H_2S, NH_3 and H_2 gaseous materials (Star et al. 2006). One of the problems associated with the decoration of CNTs with metals is to obtain metal nanoparticles of similar size well anchored to CNT sidewalls (when particles are in mobile condition on the surface, coalescence occurs and sensor response becomes unstable). This problem can be rectified by employing cold reactive plasma treatments (e.g., oxygen plasma) of CNTs (Ionescu et al. 2006, Espinosa et al. 2007). The work function of oxygen plasma-treated CNTs is very close to that of metals such as Pt, Au, Pd, Ni or Rh, making it easy and simple for electrons to move between the metal nanoparticles and the CNTs with the direction of charge transfer depending on the gaseous environment. The effective electronic interaction between metal nanoparticles and the CNT makes the process very easy in the detection of gases through the change in the electrical conductivity of these hybrid nanomaterials. This was reported by Leghrib and co-workers to selectively detect benzene molecules using arrays of plasma-treated, metal decorated multiwall carbon nanotubes (Leghrib et al. 2010, Leghrib et al. 2011, Zanolli et al. 2011).

5. Graphene as Gas Sensors

In 2007, graphene had been first reported in gas sensing (Schedin et al. 2007) in which they had demonstrated the electrical detection of gas molecules adsorbed on multiterminal Hall bars. These were fabricated in a conventional way by lithographic methods from single-layer or few-layer graphene that had been mechanically cleaved from graphite. Adsorption of gaseous materials resulted in the devices to show concentration-dependent variation in resistivity. The baseline could be readjusted by heating at 150°C under vacuum conditions. The variation in resistivity due to the gas-induction had different magnitudes for different gaseous components, and the sign in the change indicated whether the gas had been worked as an electron acceptor (e.g., nitrogen dioxide and moisture) or an electron donor (e.g., carbon monoxide, ethanol and ammonia). By considering the advantage from the low noise levels in their graphene devices, one research group had reported long-term measurements on extremely dilute NO_2 samples and had observed step-like variation in resistivity during adsorption and desorption. From statistical analyses of these quantized data, they concluded their results as evidence in the detection of adsorption or desorption of individual gas molecules. In this case, they had considered the fact that conductivity is directly proportional to the product of carrier density and mobility. Therefore, it can be affirmed that such experimental results were generated due to variation in carrier density, mobility or both. Many research groups had also reported various computational chemistry studies to analyze the theoretical concept of the adsorption of different molecules (moisture, NO_2, NO, NH_3, CO, CO_2, O_2 and N_2) on graphene (Leenaerts et al. 2008, Huang et al. 2008). Their study revealed that NO_2 acted as a strong dopant here and that moisture or NH_3 should have generated milder effects. Usually, NH_3 molecules interacted with water molecules absorbed on the devices which probably were contributing to the large response (Ratinac et al. 2008). Other researchers had also employed that conventional nano-lithography became responsible for making gas sensing devices resist residues on the graphene surface (Goldoni et al. 2003, Dan et al. 2009). This contaminated chemically-doped graphene actually increased the carrier scattering and acted as an absorbent layer which further concentrated gas molecules on the surface of graphene, resulting in the enhancement of gas response. These contaminants could be cleaned with the application of H_2/Ar and the intrinsic graphene responses to gases could be measured. These responses were found to be very small and thus suggested the requirement for surface functionalization in order to achieve sub-ppb level sensitivity. These researchers suggested the application of substitutional doping in graphene to induce its gas sensing properties (Dan et al. 2009). They had also conducted computational studies on a B, N, Al and S-substituted graphene sheet, which suggested that B or S doped graphene would be advantageous for detecting NO and NO_2. Reduced graphene oxides had also been adapted as a useful material for developing gas sensors. Graphene oxides have been proved as much easier to process than of graphene and offer the possibility to tailor the number of functional groups by controlling the degree of reduction. One research demonstrated the reduced graphene oxide (r-GO) as the active material for high-sensitive gas sensors (Robinson et al. 2008). Graphene oxide was applied to spin-coat a silicone substrate and standard photolithography was utilized to develop interdigitated arrays of Ti/Au electrodes. The processed samples were further reduced in graphene by exposing them to hydrazine hydrate vapor while being heated at 100°C. The controlling time of exposure had been considered one of the key features. These devices achieved ppb sensitivity to different warfare agent simulants. The partial reduction of graphene oxide to graphene was also required as the process left a functionalized surface with respect to active oxygen defects, which showed larger reactivity toward target analytes. However, the noise level enhanced for short reduction times and thus a trade-off exists to touch an optimal lower limit of detection. A similar approach has been also reported by other research groups in which explosive vapors were detected at ppb levels (Fowler et al. 2009). The r-GO consisted of a functionalized surface with active oxygen defects, which was reported by Li et al. to decorate RGO with Pd nanoparticles with a solution chemistry technology (Li et al. 2011). The Pd-RGO was further drop-coated onto a standard Si substrate with patterned Ni/CVD-grown graphene electrodes, and an AC dielectrophoretic technique was applied in assembling the Pd-RGO into ordered conducting channels on the hydrophobic surface of the wafer substrate. The resulting sensors showed ppb response to HNO_3.

6. Carbon Nano-fibers as Gas Sensors

Ouyang and co-workers have suggested the dispersion of carbon nanofibres within a polymer matrix as a way to overcome the instability experienced with carbon black-polymer composites (Ouyang et al. 2013). This instability occurs because the nanosized carbon black particles tend to reaggregate when the composite absorbs vapor, which lowers the matrix viscosity and increases its volume. In contrast, dispersing carbon nanofibres in the polymer improves the vapor sensing stability because these high aspect ratio fibers resist movement within their polymer composites when vapor is absorbed and desorbed. Thus, the original electrical percolation pathways present in these composites are maintained after the absorbed vapor has desorbed from the matrix. In this work, vapor grown carbon nanofibres were suspended in toluene and polystyrene was added under magnetic stirring (Ouyang et al. 2013). Lee and co-workers reported the fabrication of a polyacrylonitrile and carbon black complex as a gas sensor using electrospinning Lee et al., a method that could be easily scaled up for the mass production of inexpensive sensors. Electrospun fibers were thermally treated to obtain carbon fibers, which were then chemically activated to improve the active sites for gas adsorption. KOH solutions were the activation agents, and this process improved the porous structure, increasing the specific surface area of carbon fibers nearly 100 times. As a result, the amount of adsorbed gas is significantly increased. Additionally, the surface of activated samples was modified by a fluorination treatment. The induced functional groups help to attract the target gas to the surface of the gas sensor. The electrical conductivity was improved by the inclusion of carbon black additives. Carbon black additives help in the efficient transfer of the resistive response from the surface of the gas sensor to the electrodes. Overall, the sensor response to nitrogen oxide and carbon monoxide was improved about five times due to the effects of chemical activation, the presence of carbon black additives and the fluorination treatments. In Fong and co-workers presented a method for obtaining mats of carbon nanofibres decorated with Pd nanoparticles employing electrospinning and a wet chemistry route for surface functionalization. The material consisted of relatively uniform, randomly overlaid carbon nanofibres with diameters of about 300 nm, while the attached Pd nanoparticles had sizes in the range from a few to tens of nanometers. The surface functionalization was carried out by immersion of the mats of electrospun polyacrylonitrile fibers in a NH2OH aqueous solution. The amidoxime-functionalized mats were then immersed in a Pd(NO3)2 aqueous solution to allow amidoxime groups to coordinate/chelate with Pd2+ ions. The nano-felt was further treated in NH2OH aqueous solution to prepare the nano-mats surface attached with Pd nanoparticles. These underwent thermal treatments for the standard stabilization and carbonization processes. The same route could be employed to obtain carbon nanofibers decorated with nanoparticles of different metals. Electrospun carbon mats were mechanically flexible/resilient, and the resistance of the material varied upon exposure to hydrogen at room temperature. Sensors showed a moderate response to 100% hydrogen with a rather long response and recovery times. Reactivity strongly depends on the size of metal clusters attached to the surface of the fibers, and the method reportedly led to a rather high dispersion on Pd cluster size. Jang and co-workers reported in a method to obtain metal oxide-decorated electrospun carbon nanofibres. A remarkable advantage of the process employed is that the metal oxide-decorated carbon nanofibres were obtained on one go. To decorate metal oxides (ZnO and SnO2) on the carbon nanofibre surface, core (PAN) and shell (PVP) structured nanofibres were fabricated as starting materials. The core-shell structure was prepared by single-nozzle co-electrospinning because of the incompatibility of the two polymers. The incompatibility of the two polymers in the mixture solution resulted from their very different intrinsic physical properties, in particular their viscosities. Ultrafine hybrid CNFs were then formed by decomposing the PVP phase, converting the metal precursors (i.e., ZnAc2 and SnCl4) to metal oxide nano-nodules and transforming the PAN to CNFs of about 40 nm diameter during heat treatment. The decoration morphology of the metal oxide nano-nodules could be controlled by precursor concentration in the PVP solution. The metal oxide decorated carbon nanofibre hybrids were suspended in an ethanol solution employing ultrasonication and were subsequently deposited by spin-coating onto interdigitated electrode arrays for performing standard resistive measurements. The sensors were characterized in the presence of dimethyl methylphosphonate (DMMP) at room temperature. The minimum detectable level of hybrid carbon nanofibres was as low as 0.1 parts-per-billion (ppb). This high sensitivity is attributed to the presence of metal oxide nano-nodules on the surface of carbon nanofibres, which increase the surface area and affinity to DMMP vapors. However, how ambient humidity would affect the response is not discussed.

References

Ai, K., Y. Liu, C. Ruan, L. Lu, G.Q. Lu. 2013. Sp^2 C-dominant N-doped carbon sub-micrometer spheres with a tunable size: A versatile platform for highly efficient oxygen-reduction catalysts. *Adv. Mater.* 25, 998–1003.

Arora, N., N.N. Sharma. 2014. Arc discharge synthesis of carbon nanotubes: Comprehensive review. *Diamond Relat. Mater.* 50, 135–150.

Baccile, N., M. Antonietti, M.-M. Titirici. 2010. One-step hydrothermal synthesis of nitrogen-doped nanocarbons: Albumine directing the carbonization of glucose. *Chem. Sus. Chem.* 3, 246.

Bergius, F. 1913. Die Anwendunghoher Drückebeichemischen Vorgängen und eine Nachbildung des Entstehungsprozesses der Steinkohle, Knapp, Halle/Saale.

Bogaerts, A., Z. Chen. 2005. Effect of laser parameters on laser ablation and laser-induced plasma formation: A numerical modeling investigation. *Spectrochim. Acta B Atom Spectrosc.* 60, 1280–1307.

Bondavalli, P., P. Legagneux, D. Pribat. 2009. Carbon nanotubes based transistors as gas sensors: state of the art and critical review. *Sens. Actuators B: Chem.* 140, 304–318.

Chen, T., L. Pan, T. Lu, C. Fu, D. Chua, Z. Sun. 2014. Fast synthesis of carbon microspheres *via* a microwave-assisted reaction for sodium ion batteries. *J. Mater. Chem. A* 2, 1263–1267.

Choma, J., D. Jamioła, K. Augustynek, M. Marszewski, M. Gao, M. Jaroniec. 2012. New opportunities in Stöber synthesis: preparation of microporous and mesoporous carbon spheres. *J. Mater. Chem.* 22, 12636–12642.

Choma, J., W. Fahrenholz, D. Jamiola, J. Ludwinowicz, M. Jaroniec. 2014. Development of mesoporosity in carbon spheres obtained by Stöber method. *Microporous Mesoporous Mater.* 185, 197–203.

Collins, P.G., K. Bradley, M. Ishigami, A. Zett. 2000. Extreme oxygen sensitivity of electronic properties of carbon nanotubes. *Science* 287, 1801–1804.

Dan, Y., Y. Lu, N.J. Kybert, Z. Luo, A.T.C. Johnson. 2009. Intrinsic response of graphene vapor sensors. *Nano Lett.* 9, 1472–1475.

Demir-Cakan, R., N. Baccile, M. Antonietti, M.-M. Titirici. 2009. Carboxylate-rich carbonaceous materials via one-step hydrothermal carbonization of glucose in the presence of acrylic acid. *Chem. Mater.* 21, 484.

Demir-Cakan, R., P. Makowski, M. Antonietti, F. Goettmann, M.M. Titirici. 2010. Hydrothermal synthesis of imidazole functionalized carbon spheres and their application in catalysis. *Catal. Today* 150, 115–118.

Espinosa, E.H., R. Ionescu, C. Bittencourt, A. Felten, R. Erni, G. Van Tendeloo. 2007. Metal-decorated muliwall carbon nanotubes for low temperature gas sensing. *Thin Solid Films* 515, 8322–8327.

Falco, C., M. Sevilla, R.J. White, R. Rothe, M.-M. Titirici. 2012. Renewable nitrogen-doped hydrothermal carbons derived from microalgae. *Chem. Sus. Chem.* 5, 1834–1840.

Falco, C., J.M. Sieben, N. Brun, M. Sevilla, T.v.d. Mauelen, E. Morallón, D. Cazorla-Amorós, M.-M. Titirici. 2013. Hydrothermal carbons from hemicellulose-derived aqueous hydrolysis products as electrode materials for supercapacitors. *Chem. Sus. Chem.* 6, 374–382.

Farhat, S., M. Lamy de La Chapelle, A. Loiseau, C.D. Scott, S. Lefrant, C. Journet, P. Bernier. 2001. Diameter control of single-walled carbon nanotubes using argon–helium mixture gases. *J. Chem. Phys.* 115, 6752.

Fellinger, T.P., R.J. White, M.M. Titirici, M. Antonietti. 2012. Borax-mediated formation of carbon aerogels from glucose. *Adv. Funct. Mater.* 22, 3254–3260.

Feng, S., W. Li, Q. Shi, Y. Li, J. Chen, Y. Ling, A.M. Asirib, D.Y. Zhao. 2014. Synthesis of nitrogen-doped hollow carbon nanospheres for CO_2 capture. *Chem. Commun.* 50, 329–331.

Fowler, J.D., M.J. Aleen, V.C. Tung, Y. Yang, R.B. Kaner, B.H. Weiller. 2009. Practical chemical sensors from chemically derived graphene. *ACS Nano.* 3, 301–306.

Gamaly, E., T. Ebbesen. 1995. Mechanism of carbon nanotube formation in the arc discharge. *Phys. Rev. B, Cond. Matter.* 52, 2083–2089.

Goldoni, A., R. Larciprete, L. Petaccia, S. Lizzit. 2003. Single-wall carbon nanotube interaction with gases: sample contaminants and environmental monitoring. *J. Am. Chem. Soc.* 125, 11329–11333.

Gong, Y., Z. Wei, J. Wang, P. Zhang, H. Li, Y. Wang. 2014. Design and fabrication of hierarchically porous carbon with a template-free method. *Sci. Rep.* 4, 6349.

Gu, J., S. Su, Y. Li, Q. He, J. Shi. 2011. Hydrophilic mesoporous carbon nanoparticles as carriers for sustained release of hydrophobic anti-cancer drugs. *Chem. Commun.* 47, 2101.

Han, F., Y. Bai, R. Liu, B. Yao, Y. Qi, N. Lun, J.X. Zhang. 2011. Template-free synthesis of interconnected hollow carbon nanospheres for high-performance anode material in lithium-ion batteries. *Adv. Energy Mater.* 1, 798–801.

Heer, W., P. Poncharal, C. Berger, J. Gezo, Z. Song, J. Bettini, D. Ugarte. 2005. Liquid carbon, carbon-glass beads, and the crystallization of carbon nanotubes. *Science* 307, 907–910.

Hu, B., K. Wang, L. Wu, S.H. Yu, M. Antonietti, M.-M. Titirici. 2010. Engineering carbon materials from the hydrothermal carbonization process of biomass. *Adv. Mater.* 22, 813–828.

Huang, B., Z. Li, Z. Liu, G. Zhou, S. Hao, J. Wu, B.-L. Gu, W. Duan. 2008. Adsorption of gas molecules on graphene nanoribbons and its implication for nanoscale molecule sensor. *J. Phys. Chem. C* 112, 13442–13446.

Huang, X., S. Kim, M.S. Heo, J.E. Kim, H. Suh, I. Kim. 2013. Easy synthesis of hierarchical carbon spheres with superior capacitive performance in supercapacitors. *Langmuir* 29, 12266–12274.

Iijima, S. 1991. Helical microtubules of graphitic carbon. *Nature* 354, 56.

Ionescu, R., E.H. Espinosa, E. Sotter, E. Llobet, X. Vilanova, X. Correig, A. Felten, C. Bittencourt, G. Van Lier, J.-C. Charlier, J.J. Pireaux. 2006. Oxygen functionalisation of MWNT and their use as gas sensitive thick-film layers. *Sens. Actuators B: Chem.* 113, 36–46.

Ismail, R., A. Ali, M. Ismail, K. Hassoon. 2011. Preparation and characterization of colloidal ZnO nanoparticles using nanosecond laser ablation in water. *Appl. Nanosci.* 1, 45–49.

Ismail, R., K. Khashan, R. Mahdi. 2017. Characterization of high photosensitivity nanostructured 4H-SiC/p-Si heterostructure prepared by laser ablation of silicon in ethanol. *Mater. Sci. Semicond. Process.* 68, 252–261.

Kauffman, D.R., A. Star. 2008. Carbon nanotube gas and vapor sensors. *Angew. Chem. Int. Ed.* 47, 6550–6570.

Keidar, M. 2007. Factors affecting synthesis of single wall carbon nanotubes in arc discharge. *J. Phys. D. Appl. Phys.* 40, 2388–2393.

Kim, T., P. Chung, I.I. Slowing, M. Tsunoda, E.S. Yeung, V.S.-Y. Lin. 2008. Structurally ordered mesoporous carbon nanoparticles as transmembrane delivery vehicle in human cancer cells. *Nano Lett.* 8, 3724–3727.

Kong, J., N.R. Franklin, C. Zhou, M.G. Chapline, S. Peng, K. Cho, H. Dai. 2000. Nanotube molecular wires as chemical sensors. *Science* 287, 622–625.

Kong, J., M.G. Chapline, H. Dai. 2001. Functionalized carbon nanotubes for molecular hydrogen sensors. *Adv. Mater.* 13, 1384–1386.

Kubo, S., R.J. White, N. Yoshizawa, M. Antonietti, M.-M. Titirici. 2011. Ordered carbohydrate-derived porous carbons. *Chem. Mater.* 23, 4882–4885.

Kumar, M., Y. Ando. 2010. Chemical vapor deposition of carbon nanotubes: a review on growth mechanism and mass production. *J. Nanosci. Nanotech.* 10, 3739–3758.

Lee, J.S., O.S. Kwon, S.J. Park, E.U. Park, S.A. You, H.S. Yoon, J.S. Jang. 2011. Fabrication of ultrafine metal-oxide-decorated carbon nanofibers for DMMP sensor application. *ACS Nano* 5, 7992–8001.

Leenaerts, O., B. Partoens, F.M. Peeters. 2008. Adsorption of H_2O, NH_3, CO, NO_2 and NO on graphene: a first-principles study. *Phys. Rev. B* 77, 125416.

Leghrib, R., A. Felten, F. Demoisson, F. Reniers, J.-J. Pireaux, E. Llobet. 2010. Room-temperature, selective detection of benzene at trace levels using plasmatreated metal-decorated multiwalled carbon nanotubes. *Carbon* 48, 3477–3484.

Leghrib, R., T. Dufour, F. Demoisson, N. Claessens, F. Reniers, E. Llobet. 2011. Gas sensing properties of multiwall carbon nanotubes decorated with rhodium nanoparticles. *Sens.Actuators B: Chem.* 160, 974–980.

Li, W., X. Geng, Y. Guo, J. Rong, Y. Gong, L. Wu, X. Zhang, P. Li, J. Xu, G. Cheng, M. Sun, L. Liu. 2011. Reduced graphene oxide electrically contacted graphene sensor for highly sensitive nitric oxide detection. *ACS Nano.* 5, 6955–6961.

Liang, C., Z. Li, S. Dai. 2008. Mesoporous carbon materials: synthesis and modification. *Angew. Chem. Int. Ed.* 47, 3696–3717.

Liang, F., T. Shimizu, M. Tanaka, S. Choi, T. Watanabe. 2012. Selective preparation of polyhedral graphite particles and multi-wall carbon nanotubes by a transferred arc under atmospheric pressure. *Diamond Relat. Mater.* 30, 70–76.

Liang, H., Q. Guan, L. Chen, Z. Zhu, W. Zhang, S.-H. Yu. 2012. Macroscopic-scale template synthesis of robust carbonaceous nanofiber hydrogels and aerogels and their applications. *Angew. Chem. Int. Ed.* 51, 5101–5105.

Liu, J., S.Z. Qiao, H. Liu, J. Chen, A. Orpe, D. Zhao, G.Q. Lu. 2011. Extension of the Stöber method to the preparation of monodisperse resorcinol-formaldehyde resin polymer and carbon spheres. *Angew. Chem. Int. Ed.* 50, 5947–5951.

Ludwinowicz, J., M. Jaroniec. 2015. Potassium salt-assisted synthesis of highly microporous carbon spheres for CO_2 adsorption. *Carbon* 82, 297–303.

Ma, X., L. Gan, M. Liu, P. Tripathi, Y. Zhao, Z. Xu, D. Zhu, L. Chen. 2014. Mesoporous size controllable carbon microspheres and their electrochemical performances for supercapacitor electrodes. *J. Mater. Chem. A* 2, 8407–8415.

Mane, G.P., S.N. Talapaneni, C. Anand, S. Varghese, H. Iwai, Q. Ji, K. Ariga, T. Mori, A. Vinu. 2012. Preparation of highly ordered nitrogen-containing mesoporous carbon from a gelatin biomolecule and its excellent sensing of acetic acid. *Adv. Funct. Mater.* 22, 3596–3604.

Ouyang, Y., H. Shi, R. Fu, D. Wu. 2013. Highly monodisperse microporous polymeric and carbonaceous nanospheres with multifunctional properties. *Sci. Rep.* 3, 1430.

Ratinac, K.R., W. Yang, S.P. Ringer, F. Braet. 2010. Toward ubiquitous environmental gas sensors capitalizing on the promise of graphene. *Env. Sci. Tech.* 44, 1167–1176.

Rinzler, A.G., J. Liu, H. Dai, P. Nikolaev, C.B. Huffman, F.J. Rodriguez-Macias, P.J. Boul, A.H. Lu, D. Heymann, D.T. Colbert, R.S. Lee, J.E. Fischer, A.M. Rao, P.C. Eklund, R.E. Smalley. 1998. Large-scale purification of single-wall carbon nanotubes: process, product, and characterization. *Appl. Phys. A* 67, 29–37.

Roberts, A.D., X. Li, H. Zhang. 2014. Porous carbon spheres and monoliths: morphology control, pore size tuning and their applications as Li-ion battery anode materials. *Chem. Soc. Rev.* 43, 4341.

Roberts, M.E., M.C. LeMieux, Z. Bao. 2009. Sorted and aligned single-walled carbon nanotube networks for transistor-based aqueous chemical sensors. *ACS Nano.* 3, 3287–3293.

Robinson, J.T., F.K. Perkins, E.S. Snow, Z. Wei, P.E. Sheehan. 2008. Reduced graphene oxide molecular sensors. *Nano Lett.* 8, 3137–3140.

Román, S., J.M. Valente Nabais, B. Ledesma, J.F. González, C. Laginhas, M.M. Titirici. 2013. Production of low-cost adsorbents with tunable surface chemistry by conjunction of hydrothermal carbonization and activation processes. *Microporous Mesoporous Mater.* 165, 127–133.

Schedin, F., A.K. Geim, S.V. Morozov, E.W. Hill, P. Blake, M.I. Katsnelson, K.S. Novoselov. 2007. Detection of individual gas molecules adsorbed on graphene. *Nat. Mater.* 6, 652–655.

Schultzenberger P., L. Schultzenberger. 1890. Sur quelques faits relatifs à l'histoire du carbone. *C.R. Acad. Sci. Paris* 111, 774–778.

Sevilla, M. and A.B. Fuertes. 2009. The production of carbon materials by hydrothermal carbonization of cellulose. *Carbon* 47, 2281–2289.

Sevilla, M., A.B. Fuertesa, R. Mokaya. 2011. High density hydrogen storage in superactivated carbons from hydrothermally carbonized renewable organic materials. *Energy Environ. Sci.* 4, 1400–1410.

Sevilla, M., C. Falco, M.-M. Titiricic, A.B. Fuertes. 2012. High-performance CO_2 sorbents from algae. *RSC Adv.* 2, 12792–12797.

Song, J.C., Z.Y. Lu and Z.Y. Sun. 2014. A facile method of synthesizing uniform resin colloidal and microporous carbon spheres with high nitrogen content. *J. Colloid Interface Sci.* 431, 132–138.

Star, A., V. Joshi, S. Skarupo, D. Thomas, J.C.P. Gabriel. 2006. Gas sensor array based metal-decorated carbon nanotubes. *J. Phys. Chem. B* 110, 21014–21020.

Stöber, W., A. Fink, E.J. Bohn. 1968. Controlled growth of monodisperse silica spheres in the micron size range. *J. Colloid Interface Sci.* 26, 62–69.

Tan, S., A. Verschueren, C. Dekker. 1998. Room-temperature transistor based on a single carbon nanotube. *Nature* 393, 49–52.

Tanaka, S., H. Nakao, T. Mukai, Y. Katayama, Y. Miyake. 2012. An experimental investigation of the ion storage/transfer behavior in an electrical double-layer capacitor by using monodisperse carbon spheres with microporous structure. *J. Phys. Chem. C* 116, 26791–26799.

Tang, K., L. Fu, R.J. White, L. Yu, M.M. Ttirici, M. Antonietti, J. Maier. 2012. Hollow carbon nanospheres with superior rate capability for sodium-based batteries. *Adv. Energy Mater.* 2, 873–877.

Titirici, M.M., A. Thomas, M. Antonietti. 2007. Replication and coating of silica templates by hydrothermal carbonization. *Adv. Funct. Mater.* 17, 1010–1018.

Titirici, M.M., A. Thomas, S.-H. Yu, J.-O. Müller, M. Antonietti. 2007. A direct synthesis of mesoporous carbons with bicontinuous pore morphology from crude plant material by hydrothermal carbonization. *Chem. Mater.* 19, 4205–4212.

Titirici, M.M., R.J. White, C. Falco, M. Sevilla. 2012. Black perspectives for a green future: hydrothermal carbons for environment protection and energy storage. *Energy Environ. Sci.* 5, 6796–6822.

Ugarte, D. 1994. High-temperature behaviour of 'fullerene black'. *Carbon N. Y.* 32, 1245–1248.

Unur, E. 2013. Functional nanoporous carbons from hydrothermally treated biomass for environmental purification. *Microporous Mesoporous Mater.* 168, 92–101.

Wang, L., Q. Sun, X. Wang, T. Wen, J.-J. Yin, P. Wang, R. Bai, X.-Q. Zhang, L.-H. Zhang, A.-H. Lu, C. Chen. 2015. Using hollow carbon nanospheres as a light-induced free radical generator to overcome chemotherapy resistance. *J. Am. Chem. Soc.* 137, 1947–1955.

Wei, J., D. Zhou, Z. Sun, Y. Deng, Y. Xia, D.Y. Zhao. 2013. A controllable synthesis of rich nitrogen-doped ordered mesoporous carbon for CO_2 capture and supercapacitors. *Adv.Funct. Mater.* 23, 2322–2328.

White, R.J., K. Tauer, M. Antonietti, M.-M. Titirici. 2010. Functional hollow carbon nanospheres by latex templating. *J. Am. Chem. Soc.* 132, 17360–17363.

White, R.J., N. Yoshizawa, M. Antonietti, M.-M. Titirici. 2011. A sustainable synthesis of nitrogen-doped carbon aerogels. *Green Chem.* 13, 2428.

White, R.J., N. Brun, V. Budarin, J.H. Clark, M.M. Titirici. 2014. Always look on the "light" side of life: Sustainable carbon aerogels. *Chem. Sus. Chem.* 7, 670–689.

Wickramaratne, N.P., M. Jaroniec. 2013. Activated carbon spheres for CO_2 adsorption. *ACS Appl. Mater. Interfaces* 5, 1849–1855.

Wickramaratne, N.P. and M. Jaroniec. 2014. Tailoring microporosity and nitrogen content in carbons for achieving high uptake of CO_2 at ambient conditions. *Adsorption*, 20, 287–293.

Wu, Y., T. Zhang, F. Zhang, Y. Wang, Y. Ma, Y. Huang, Y. Liu, Y. Chen. 2012. *In situ* synthesis of graphene/single-walled carbon nanotube hybrid material by arc-discharge and its application in supercapacitors. *Nano Energy* 1, 820–827.

Xie, Z., R. White, J. Weber, A. Tauberti, M.M. Titirici. 2011. Hierarchical porous carbonaceous materials *via* ionothermal carbonization of carbohydrates. *J. Mater. Chem.* 21, 7434–7442.

Xu, S., Y. Luo, B. Tan. 2013. Recent development of hypercrosslinked microporous organic polymers. *Macromol. Rapid Commun.* 25, 471–484.

Yang, T., J. Liu, R. Zhou, Z. Chen, H. Xu, S.Z. Qiao and M.J. Monteiro. 2014. N-doped mesoporous carbon spheres as the oxygen reduction reaction catalysts. *J. Mater. Chem. A* 2, 18139–18146.

You, L.J., S. Xu, W.F. Ma, D. Li, Y.T. Zhang, J. Guo, J.J. Hu, C.C. Wang. 2012. Ultrafast hydrothermal synthesis of high quality magnetic core phenol–formaldehyde shell composite microspheres using the microwave method. *Langmuir* 28, 10565–10572.

Yu, L., C. Falco, J. Weber, R. White, J. Howe, M. Titirici. 2012. Carbohydrate-derived hydrothermal carbons: a thorough characterization study. *Langmuir* 28, 12373–12383.

Zanolli, Z., R. Leghrib, A. Felten, J.-J. Pireaux, E. Llobet, J.-C. Charlier. 2011. Gas sensing with Au-decorated carbon nanotubes. *ACS Nano.* 5, 4592–4599.

Zhang, C., K. Hatzell, M. Boota, B. Dyatkin, M. Beidaghi, D. Long, W. Qiao, E. Kumbur, Y. Gogotsi. 2014. Highly porous carbon spheres for electrochemical capacitors and capacitive flowable suspension electrodes. *Carbon* 77, 155–164.

Zhang, P.F., J.Y. Yuan, H.R. Li, X.F. Liu, X. Xu, M. Antonietti, Y. Wang. 2013. Mesoporous nitrogen-doped carbon for copper-mediated Ullmann-type C–O/–N/–S cross-coupling reactions. *RSC Adv.* 3, 1890–1895.

Zhang, P.F., J. Yuan, T. Fellinger, M. Antonietti, H.R. Li, Y. Wang. 2013. Improving hydrothermal carbonization by using poly(ionic liquid)s. *Angew. Chem. Int. Ed.* 52, 6028–6032.

Zhang, P.F., Y. Gong, H.R. Li, Z. Chen, Y. Wang. 2013. Solvent-free aerobic oxidation of hydrocarbons and alcohols with Pd@N-doped carbon from glucose. *Nat. Commun.* 4, 1593.

Zhang, P.F., Y. Gong, Z. Wei, J. Wang, H. Li, Y. Wang. 2014. Updating biomass into functional carbon material in ionothermal manner. *ACS Appl. Mater. Interfaces* 6, 12515–12522.

Zhang, P.F., Z. Qiao, Z. Zhang, S. Wan, S. Dai. 2014. Mesoporous graphene-like carbon sheet: high-power supercapacitor and outstanding catalyst support. *J. Mater. Chem. A* 2, 12262–12269.

Zhao, J., W. Niu, L. Zhang, H. Cai, M. Han, Y. Yuan, S. Majeed, S. Anjum, G. Xu. 2013. A template-free and surfactant-free method for high-yield synthesis of highly monodisperse 3-aminophenol–formaldehyde resin and carbon nano/microspheres. *Macromolecules* 46, 140–145.

Zhao, L., L.-Z. Fan, M.-Q. Zhou, H. Guan, S. Qiao, M. Antonietti, M.-M. Titirici. 2010. Nitrogen-containing hydrothermal carbons with superior performance in supercapacitors. *Adv. Mater.* 22, 5202–5206.

Zhao, Q., M.B. Nardelli, W. Lu, J. Bernhoc. 2005. Carbon nanotubesmetal cluster composites: a new road to chemical sensors. *Nano Lett.* 5, 847–851.

Zhou, D., L. Chow. 2003. Complex structure of carbon nanotubes and their implications for formation mechanism. *J. Appl. Phys.* 93, 9972.

Chapter **6**

Nanomaterials Based Biosensor

1. Introduction

A biosensor as delineated by IUPAC is a device that uses precise biochemical reactions mediated by isolated enzymes, immunosystems, tissues, organelles or whole cells to detect chemical compounds, usually by electrical, thermal or optical signals (Mcnaught and Wilkinson 1997). The elemental idea of using a biosensor was reported first by Clark and his research group in 1953 through his published paper. From then on the development of assorted types of biosensors started, and now they are widely used in medical, forensic science and environmental science fields for the detection of pollutants. The basic function accomplished by biosensors is to sense biomaterials like antibodies, proteins, enzymes and many more, and recognition at the molecular level is the fundamental prerequisite for that.

The key basic component of the biosensor are bioreceptors, transducers and the detector based on the fundamental mode of operation of the Biosensor (Figure 6.1). Specificity and selectivity depend on a recognition system for biomolecules connected with transducer (Turner et al. 1987, Kricka 1988, Buch and Rechnitz 1989).

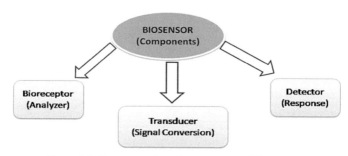

Figure 6.1: Representation of components of biosensors.

➢ **Bioreceptor**

This component acts as a template for materials to be analyzed. A broad range of materials are used as receptors, such as an antibody can be analyzed using an antigen and a protein can be detected using its analogous substrate.

➢ **Transducers**

The transducer transforms the biochemical interaction between bioreactors and the bioanalyte in form of an electrical signal which is perceptible.

➢ **Detector**

The third important component of the biosensor is to collect the electrical signal received from the transducer and after an appropriate amplification, the data is used for qualitative and quantitative determination (Malik et al. 2013).

In the present time, nanotechnology and the application of nanomaterials have shown a blooming development which has advanced greatly with the gradual discovery of novel properties of nanomaterials. After the introduction of nanotechnology in the field of biosensing, the preciseness and specificity of the biosensor for biomolecule detection have been markedly upgraded. Due to their very small size measuring 1–100 nm (Buzea and Pacheco 2007), they have the efficiency to detect or manipulate atoms and molecules, thus have found their way in the field of biomolecule detection, medical diagnosis and monitoring environment (Zhang et al. 2008, Cui 2007, Pan et al. 2008).

2. Nanomaterial-Based Biosensors

Various nanomaterials have assisted in the development of a highly specific biosensor. The unification of nanotechnology with biosensing technology has encouraged the development of a new field:nanobiosensors. The miniaturized nanoscale size of nanomaterials make them a very influential material with exceptional physicochemical properties and having their atoms located adjacent to their surface has also proved very effective for the sensing mechanism (Pandit 2016). Owing to these properties, several nanomaterials have been scrutinized for their application in an improved biological signaling and effective transduction mechanism.

Some of the nanomaterials diversely employed in the advancement of biosensors are nanoparticles, nanorods, nanotubes, quantum dots and graphene.

(a) Nanoparticles-Based Biosensor

Metal-based nanoparticles have entirely transformed the field of biosensing and are efficiently used for biosensing devices implemented either in transducers or receptor part of biosensor. Diversified nanoparticles have been utilized for the construction of biosensor, such as gold nanoparticles, silver nanoparticles, semiconductor nanoparticles and silver-silica hybrid nanoparticles (Cai et al. 2002, Xiao et al. 2003, Schierhann et al. 2006). These diverse forms of nanoparticles play a distinct role in the biosensing system. Among all, gold nanoparticles (AuNPs) have received the greatest interest because of their various fascinating properties (Wang et al. 2001).

(i) Gold Nanoparticles-Based Biosensors (AuNPs-Biosensor)

Gold-based biosensors can be classified as optical biosensors, electrochemical biosensors and piezoelectric biosensors. The role played by each type of AuNPs in the biosensing system is presented in Table 6.1.

Table 6.1: Different role played by AuNPs in biosensing system (Source: Li et al. 2010).

Types of Biosensor	Principle of Detection	Functions of AuNPs	Properties Used	Sensor Advantage
Optical biosensor	Changes in optical properties	Refractive index changes enhancement	Large dielectric constant, High molecular weight and density	Improved sensitivity
		Electron transfer rate enhancement	Conductivity, Quantum dimension	Improved sensitivity
Electrochemical biosensor	Changes in electrical characteristics	Immobilization platform	Biocompatibility, Large surface area	Improved sensitivity and stability
		Catalysis of reactions	High surface energy	Improved sensitivity and selectivity
Piezoelectric biosensor	Changes in mass	Biomolecule immobilization	Biocompatibility, High density, Large surface-to-volume ratio	Improved sensitivity

➤ **AuNP-based Optical Biosensor**

   An optical biosensor device detects and analyzes changes in photon output. The collective oscillations of electrons present in conduction (Plasmons) in return for external electromagnetic radiation provide a unique optical property to AuNPs (Murphy 2008). At the time of interaction between electromagnetic radiation and conduction, electron gives rise to a phenomenon known as Surface Plasmon Resonance (SPR) used for distinguishing any physiochemical changes of thin-film present on the surface of the metal (Daniel and Astruc 2004, Chen et al. 2006). Any molecule which attaches specifically to the surface of these metallic surface films induces a change in the reflection pattern of laser light from the metal-liquid surface due to deviation in dielectric constant (Figure 6.2). AuNPs which show a sensitive optical extinction spectrum with the dielectric constant of the medium has captivated intensive research worldwide (Link and El-Sayed 1996, Mulvaney 1996).

Reportedly, AuNPs amplify SPR signal to a certain extent (Lyon et al. 1998, Lyon et al. 1999, He et al. 2000, Hutter et al. 2001, Lin et al. 2006, Takae et al. 2007, Fu et al. 2007, Cao and Sim 2007, Xu et al. 2009). In 2006, Lin et al. devised a fiber-based biosensor for the detection of organophosphorus pesticides using SPR effects of AuNPs (Lin et al. 2006). On AuNPs' layer, another layer of acetylcholinesterase (AChE) was immobilized and when the specific pesticide was presented AChE mediated hydrolysis of acetylcholine is inhibited. This inhibition brings about a change in the light attenuation and a correlative analysis between the rate of inhibition and light attenuation concentration of pesticide could be dogged (Lin et al. 2006). Studies have suggested that AuNPs coated on optical fiber enhance the sensitivity of biosensors substantially.

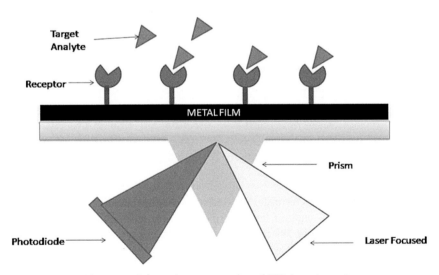

Figure 6.2: Schematic representation of SPR detection unit.

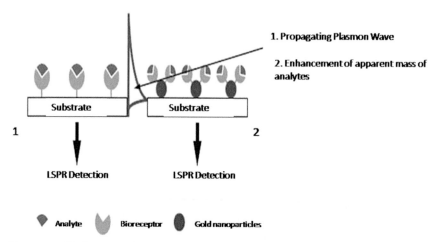

Figure 6.3: SPR biosensor (1) Interaction between localized surface plasmon of AuNPs and propagating Plasmon on SPR surface (2) AuNPs increase mass of analytes immobilized (Reproduced with permission from Li et al. 2010).

A research group headed by Li in 2006 developed a method (Figure 6.3) for analysis of single nucleotide polymorphism (SNPs) in DNA using Surface Plasmon Resonance Imaging merged with AuNPs (Li et al. 2006). Based on AuNPs, the SPR Biosensor chip was developed by Matsui and group for the detection of a molecule like dopamine which is very small in size (Matsui et al. 2005). The miniaturized nanoparticles can also be incorporated into other optical structures. Very recently AuNPs based immune sensor was reported by Tseng and group based on fiber-optic interferometry (Tseng et al. 2008).

➢ **AuNP-Based Electrochemical Biosensor**

Electrochemical Biosensor can be delineated as a device that converts biomolecule interacting phase into an electrical signal. It has the ability to provide rapid and cost-effective detection and has received major attention from researchers. After the application of AuNPs sensitivity as well as specificity of the electrochemical biosensor can be enhanced.

i. **AuNP as Electrochemical Indicators**—Redox reaction between Au^0 and Au^{3+} is the base of using AuNPs as an electrochemical indicator (Garcia and Garcia 1995). AuNPs signal in electrochemical biosensor can be detected by any of the methods discussed below.

• The oxidation signal of AuNPs is detected directly without any treatment (Ozsoz et al. 2003). An electrochemical sensor was reported for DNA detection ensuring the method of direct oxidation of AuNPs and provided a detection limit of 2.17 pM. The main restraint of this method is that majority of AuNPs cannot be detected as their distance from the electrode is much more.

- AuNP is electro-oxidized to gold ions in the presence of HCl (Hydrochloric acid solution (Pumera et al. 2005, Zheng et al. 2007, Ambrosi et al. 2007, Afonso et al. 2013, Lau et al. 2017, Wang and Alocilja 2015). To reduce this limitation of the direct oxidation method, AuNP is oxidized to gold ions for more precise detection. This method was reported by Trau and group where AuNPs were oxidized to Au^{3+} by using HCl solution and the reduction of gold ions was detected. In this electrochemical sensor, a detection limit as low as 1 colony-forming unit (CFU) was observed for DNA of Mycobacterium tuberculosis (Ng et al. 2015).

- A signal is detected after the dissolution of AuNPs in acidic solutions. Sensor-based on this method was devised by Ilkhani and group in which gold preconcentration at the cathodic end and gold stripping at the anode end were performed after dissolving AuNPs in HCl solution. The detection limit obtained for this biosensor was 50 pg.mL^{-1} (Ilkhani et al. 2015). Gold ion preconcentration at the cathodic end enhances the electrochemical signal.

  Qin et al. also reported an electrochemical biosensor based on this method where at the cathodic end a cathodic potential (0 V) was applied and then the dissolution of AuNPs and gold ion preconcentration at the cathode was concurrently performed in microliter-droplet aqueous HBr/Br_2. Signal recovery by this scheme was much higher and the estimation limit recorded was 0.3 fg.mL^{-1}–0.1 fg.mL^{-1} (Qin et al. 2015).

  Generally, the use of corrosive acid HCl and HBr/Br_2 for the dissolution of AuNPs in electrochemical biosensors is harmful to both environment and human health. So many green reagents which have reportedly shown better electro-oxidation performance like $NaNO_3/NaCl$ has been proposed to replace acidic electrolyte (López-Marzo et al. 2018).

ii. **AuNPs as Electron Migration Enhancers**—Redox reaction taking place in an electrochemical biosensor causes the exchange of electron on the electrode with respect to the analyte concentration. When we opt for direct electrochemical detection, the result is not promising because the biomolecule shows a weak electrical conductivity which is not sufficient for the transfer of electrons to the electrode. To enhance the electrical conductivity, AuNPs are immobilized on the electrode surface which intensifies the rate of electron transfer and also enlarges the sensing area. It was in 1996 when Natan led research group demonstrated direct electron transfer of AuNPs between protein and electrode (Brown et al. 1996), and since then many research in the field of AuNPs use an enhancer of electron migration (Heydari-Bafrooei and Shamszadeh 2017, Wang et al. 2018, Zhang et al. 2018, Vural et al. 2018, Del Caño et al. 2018). Various methods are used to immobilize AuNPs on electrode surfaces, like electrodeposition, electrochemical deposition and direct immobilization (Zhao and Ma 2017, Bao et al. 2018). Size of AuNPs also influence the performance of electrochemical biosensor and affect the output signal produced by the assay. To immobilize more AuNPs on the surface of electrode a 3-D structure of AuNPs was developed by Wang and group. The scheme consists of layer by layer assembly of AuNPs on the electrode surface by using para-sulfonatocalix[4]arene (pSC$_4$) modified, AuNPs and HMD(1,6-hexanediamine) conjugation (Figure 6.4). As a result, the migration rate of the electron got intensified along with an enormous surface area. The detection limitation showed by this scheme was 0.5 ng.mL^{-1} (Wang et al. 2018).

To further intensify the conductivity of electrochemical biosensors, AuNPs merged with other highly conductive materials that are also used listed in Table 6.2.

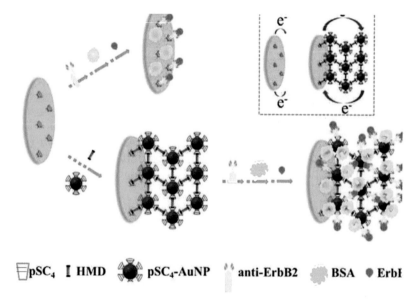

☐pSC$_4$ ∎ HMD   pSC$_4$-AuNP   anti-ErbB2   BSA   ● ErbI

Figure 6.4: Schematic representation of pSC4 monolayer and modified pSC4-gold nanoparticles layer by layer (reproduced with permission from Wang et al. 2018, Jiang et al. 2018).

Table 6.2: AuNPs-based electrochemical sensor coupled with other materials.

Type of Sensor	Materials Used	Detection Limit	Reference
Electrochemical DNA sensor	Chitosan-graphene and Polyaniline	2.11 pM	Wang et al. 2014
Electrochemical Immunosensor	AuNPS and Nile Blue A	1 pg.mL^{-1}	Gao et al. 2016
Ultrasensitive Electrochemical Sensor	AuNPs and Tungsten oxide-Graphene composite	0.05 fM	Shuai et al. 2016
Electrochemical Sensor	Carbon nanotubes and AuNPs	8 pM–11 pM	Bai et al. 2012
Electrochemical Sensor	AuNPs and Graphitized mesoporous carbon nanoparticles	0.25 ng.mL^{-1}	Liu et al. 2012
Electrochemical Sensor	Dendrimer encapsulated AuNPs	4.4 pg.mL^{-1}	Jeong et al. 2013
Electro Immunosensor	PAMAM-AuNPs	50 CFU.mL^{-1}	Zhang et al. 2016

iii. **AuNPs as Immobilization Platform**—Electrochemical biosensors are highly sensitive which depends on the number of the electroactive molecule. The large surface area and easy binding ability of AuNPs with biomolecules help in there use in immobilization platform for detection of a large number of biomolecules (Wang et al. 2003, Hu et al. 2008, Zhao et al. 2011, Yin et al. 2012, Liu et al. 2014, Wang et al. 2014, Zong et al. 2016, Zhou et al. 2016). An electrochemical sensor was developed in 2014 where a large number of DNA probes labeled with methylene blue were immobilized on the AuNPs surface. The extensive surface area of AuNPs helped in intensifying the electrochemical signal of Methylene Blue with a detection limit of 50 fM (Wang et al. 2014). The 6-ferrocenyl hexanethiol(Fc) and aptamers were immobilized simultaneously on the AuNPs surface by Shu and group. The electrochemical signal of 6-ferrocenyl hexanethiol not only got intensified but the detection sensitivity of the biosensor also increased (Shu et al. 2013).

The utilization of AuNPs in form of a platform for immobilization of DNA to detect Ampicillin was reported by Wang and his research group (Wang et al. 2018). AuNPs are often combined with other nanomaterials to improve their activity as a biosensor.

iv. **AuNPs as Catalyst**—AuNPs show some extraordinary catalytic potentiality in comparison to chemically inert bulk gold (Valden et al. 1998, Lou et al. 2001, El-Deab et al. 2003, Dela Escosure et al. 2016). The catalytic property owned by AuNPs is due to its size miniaturization, the surface-to-volume ratio and its ability to reduce over potential electrochemical reaction due to interface-dominated properties. These all bring out the enhanced detection sensitivity.

AuNPs are utilized to catalyze some important redox reactions like NADH (Nicotinamide adenine dinucleotide), H_2O_2 (Hydrogen peroxide), catechol, nitrite, etc. (Rao et al. 2017, Qu et al. 2018, Zheng et al. 2018). An immunosuppressor scheme for alpha-fetoprotein (AFP) analysis was planned by Li and group in which magnetic carbon nanotube with AuNPs was utilized whose prime function was the absorption of lead and antibodies (Li et al. 2015).

Application of GLSS (gold-labeled silver stain) in electrochemical biosensor is for intensifying electrochemical signal (Wang et al. 2001, Cai et al. 2002, Wang et al. 2002, Cai et al. 2002, Min et al. 2010, Lin et al. 2011, Lin et al. 2012, Pan et al. 2014) and AuNPs quantity directly influence detection sensitivity of GLSS. Varied electrochemical biosensors based on AuNPs are summarized in Table 6.3.

➢ **Piezoelectric Biosensor**

A very exclusive property exhibited by Quartz crystal is that when a mechanical force is enforced on Quartz crystal it generates electric potential in direction of force applied, but when an electrical field is enforced on the quartz crystal it generates mechanical vibration (Curie and Curie 1880). If any substance gets adsorbed on the Quartz crystal surface, there is a detectable shift in resonance frequency. This mass change on surface of quartz crystal is detectable by analyzing frequency shift according to piezoelectric effect and based on this effect a Quartz crystal microbalance (QCM) biosensor was designed, which is used in the medical research field for detecting genes, cells, proteins, any microorganism and toxic substances (Wang et al. 2008, Fonseca et al. 2011, Zhang et al. 2014, Zhao et al. 2015, Akter et al. 2015, Deng et al. 2016, Zhou et al. 2016).

To revise the detection sensitivity of QCM biosensors, AuNPs are often used in an immobilized form on the surface of quartz crystal (Zhao et al. 2001). It was also reported that QCM biosensor coated with platinum along with AuNPs provide maximum immobilization amount of HS-DNA (Liu et al. 2005). AuNPs plays a vital role in the amplification of signal in QCM biosensor as they are heavier in mass than any other biomolecules and also due to their higher density AuNPs increases mass change on the surface of quartz crystal (Liu et al. 2003, Kim et al. 2007, Yan et al. 2015). AuNPs of 50 nm size were used in an experimental design by Jiang and the group to amplify the

Table 6.3: AuNPs-based electrochemical biosensor (Source: Jiang et al. 2018).

Analytes[1]	Electrode Modification[2]	Functions of AuNPs	Detection Limit
Mtb DNA	SPCE/SA	Electrochemical indicators	1 CFU
EGFR	GCE	Electrochemical indicators	50 pg/mL
hIgG, hPSA	GCE/MWCNT/AB	Electrochemical indicators	0.3 fg/mL, 0.1 fg/mL
hMMP9	SPCE/AB	Electrochemical indicators	0.06 ng/mL
PSA	GCE/AuNPs/AB	Electron migration enhancers	145.69 fg/mL
M.SssIMTase	GCE/AuNPs/CP	Electron migration enhancers	0.04 U/mL
ErbB2	GE/pSC4/HMD/AuNPs	Electron migration enhancers	0.5 ng/mL
ssDNA	GCE/CS-GS/PANI/AuNPs/CP	Electron migration enhancers	2.11 pM
CEA	GCE/NB-ERGO/AuNPs/AB	Electron migration enhancers	1 pg/mL
MicroRNA	GCE/WO3-Gr/AuNPs/CP	Electron migration enhancers	0.05 fM
ssDNA	GE/CP	Immobilization platform	50 fM
Ampicillin	GCE/AuNPs/Aptamer	Immobilization platform	0.3 pM
TB	GCE/AuNPs/Aptamer	Immobilization platform	23 fM
MiRNA	GE/CP	Immobilization platform	6.8 aM
AFP	GCE/AuNPs/AB	Catalyst	3.33 fg/mL
Lysozyme	GE/CP	Catalyst	0.32 pM
Microcystin-LR	GCE/CNT/PEG	Catalyst	0.1 ng/L
PSA	GCE/Au@N-GQDs/AB	Catalyst	0.003 pg/mL

Abbreviation Used- 1. MtbDNA: Mycobacterium tuberculosis DNA; EGFR: epidermal growth factor receptor; hIgG: human immunoglobulin G; hPSA: human prostate-specific antigen; hMMP9: human matrix metallopeptidase-9; PSA: prostate-specific antigen; M.SssIMTase: methyltransferase; ErbB2: human epidermal growth factor receptor 2; CEA: carcinoembryonic antigen; TB: thrombin; AFP: alpha fetoprotein

2. SPCE: screen-printed carbon electrode; SA: streptavidin; GCE: glassy carbon electrode; MWCNT: multiwalled carbon nanotube; AB: antibody; GE: gold electrode; pSC4: para-Sulfonatocalix[4]arene; HMD: 1,6-hexanediamine; CS-GS: chitosan-graphene sheets; PANI: polyaniline; CP: capture probe; NB-ERGO: Nile blue A (NB) hybridized electrochemically reduced graphene oxide; WO3: tungsten oxide; PGE: pyrolytic graphite electrode; CNT: Carbon nanotubes; PEG: polyethylene glycol; Au@N-GQDs: AuNPs functionalized nitrogen-doped graphene quantum dots.

signal frequency of QCM, and the detection limit recorded was around 10^{-14} M for DNA (Zhao et al. 2001). This detection limit was modified to 10^{-16} M by some modifier AuNPs on the electrode surface and labeling AuNPs with probe DNA simultaneously (Liu et al. 2002, Liu et al. 2004). Reportedly using antibodies modified with AuNPs signal strength can be intensified around 53% (Kim et al. 2010). An advanced displacement type QCM biosensor was proposed in 2013 based on AuNPs, where the low detection limit achieved for Brevetoxin-B was 0.6 pg.mL^{-1}. The gold-labeled silver stain (GLSS) technique also enhances mass change in the QCM biosensor. In this scheme, AuNPs act as a catalyst and brings about a reduction of silver ions to silver element in presence of the reducing agent hydroquinone (HQ). The silver element formed gets accumulated on the AuNPs surface which increases mass change and enhances detection ability.

It has been observed that detection sensitivity can also be increased if AuNPs are combined with other biological amplifying technologies. AuNPs and Hybridized Chain Reaction (HCR) can be combined for multicycle signal amplification. The detection limit recorded from this scheme for target DNA was low as 0.7 fM (Song et al. 2018).

➢ ICP-MS Biosensor (Induced Coupled Plasma Mass Spectrometry)

The inductively coupled plasma has ionization attributes at high temperature and mass. The spectrometer offers a very fast scanning of any element, isotope and morphological analysis (Houk et al. 1980). Both these properties are coupled in this biosensor, and a very low estimation limit is achieved and widely used in the field of environmental science, biological and medical field, material analysis and many other fields (Jenner et al. 1990, Yuan et al. 2004, Jackson et al. 2004, Becker et al. 2010). AuNPs can be used in ICP-MS biosensor to achieve a high sensitivity level for the detection of biomolecules. An ICP-MS biosensor was developed by He and his research group in 2014 for the detection of HIV-1 p24 antigen detection based on AuNPs. In the model, diluted HNO_3 was used for dissociation of AuNPs from the immunoassay complex (Figure 6.5). From time to time modification has been suggested by various scientists and one such modification was suggested by Yang with his research team after an experimental trial. In his model, Yang used layer-by-layer assembly of AuNPs which was employed to amplify the ICP-MS signal for more precise detection of cancerous cells, and the detection limit recorded was 100 cell mL^{-1} (Yang et al. 2016). ICP-MS is mostly used by researchers for analysis of concentration and composition of elements. It is also capable to analyze comfortably structure, shape, particles size and

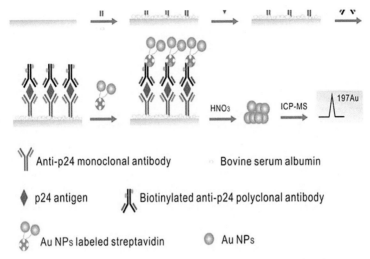

Anti-p24 monoclonal antibody   Bovine serum albumin

p24 antigen   Biotinylated anti-p24 polyclonal antibody

Au NPs labeled streptavidin   Au NPs

Figure 6.5: Schematic representation of AuNPs based immunoassay for determination of p24 antigen by ICP-MS (Reproduced with permission from He et al. 2014, Jiang et al. 2018).

with the use of AuNPs in ICP-MS biosensor its application field has also widened, and now it can be used in a biological assay (Degueldre et al. 2004, 2006, Helfrich et al. 2006, Beermann et al. 2007, Bao et al. 2016).

(b) Carbon Nanostructure Based Biosensor

Carbon miniaturized at the nanoscale range has several exclusive properties and are widely utilized as an electrochemical transducer in biosensor (Valentini et al. 2013) in any of the form discussed below.

(i) Carbon nanotubes (CNTs)—Discovered in the early 1990s and since then owing to their extraordinary properties, it has attracted the attention of various researchers. Some of the exclusive properties of nanotubes are excellent electrical conductivity, geometric features which are flexible, mechanical strength and the most important one is its dynamic physiochemical properties. Carbon nanotubes have been used in the designing of glucose biosensors. In this type of biosensor, nanotubes are used as immobilizing surfaces for glucose oxidase enzyme which is used for the estimation of glucose present in body tissues. With the application of nanotubes in the biosensor, it is now possible to estimate glucose even from body fluids like saliva and tears (Azamian et al. 2002) because nanotubes have high enzyme loading capacity and electrical conductivity is also high. Nanotubes

have also effectively been utilized in improving the performance of catalytic biosensors with enhanced oxidoreductase activity shown in both glucose oxidase and flavin adenine dinucleotide (Elie et al. 2002). CNTs coupled with sensing molecules of biosensor improve their chemo-electroluminescence effect.

(ii) **Nanowire**—They are the cylindrical structure of length ranging from few micrometers to centimeters and diameter within the nanosize range. Nanowire shows an excellent electron-transport property and is utilized for the detection of biomolecules. Biosensor based on silicon nanowires and doped with Boron has been used for the detection of biomolecules and other chemical compounds (Cui and Lieber 2001). Likewise, these silicon nanowires coated with biotin have been used for streptavidin molecule detection. Owing to their nano-size, they are also used for the detection of pathogens. Wang and the team used optical fiber of nanosize diameter and decorated with antibodies to detect toxic molecules present in cells (Wang et al. 2002).

Nanowire has proved more versatile in performance than nanotubes yet biosensors based on nanowire are less common in application. As reported by various researchers in nanowire-based biosensors, it is difficult to overall improve the electrical conductivity performance (Umar et al. 2009).

(c) Quantum Dots (QDs)

Another very promising nanomaterial used in the biosensor is luminescent semiconducting nanocrystals known as Quantum Dots (QDs). These crystals possess extraordinary photophysical properties and comprise elements belonging to the periodic group II–VI, III–V or IV–VI. The most typically used QDs are Cd-Chalcogenide nanocrystal of size about 15 cm whose shell is made of ZnS and core region composed of the centrosome (Algar et al. 2010). Photophysical properties exhibited by QDs are very unique and not found in any other material. They also provide a large absorption spectrum which is size-based (Murray et al. 1993, Park et al. 2007). This unique property of QDs is attributed to varying band gaps found in the semiconductor which is different for different nanocrystal sizes leading to the emission of distinct wavelengths after a combination of the electron-hole exciton (Weller 1993). Thus, QDs can be efficiently utilized in biosensors (Gibler et al. 2010) and currently been a subject of intense investigation.

Various QDs based biosensors have been designed, viz., QD-based FRET (Fluorescence Resonance Energy Transfer), QD-based FRET immunosensor and QD-based BRET (Bioluminescence Resonance Energy

Transfer) (Jares-Erijman et al. 2003, Huang et al. 2006). QD-based FRET genosensor is widely used in biological studies. It was reported that green and orange CdTe Quantum dots can be used as a fluorescent probe, which is pH sensitive to monitor proton flux propelled by the synthesis of ATP for the detection of viruses.

(d) Graphene and Graphene-based Nanomaterials as Biosensor

Graphene-based nanomaterials have been employed in designing various biosensor with enhanced performance. Graphene-based nanomaterials demonstrate an enhanced signal response, possess high surface area, high electron transfer rate and are biocompatible with a variety of biomolecules (Morales-Narvaez et al. 2017, Chauhan et al. 2017, Janegitz et al. 2017). Graphene-based nanomaterials are used as transducers in biosensors because they can expertly convert the signal generated by the synergy between the receptor and target molecule which is easily detectable (Pumera et al. 2011). It was reported that Graphene-based nanomaterial facilitates a high rate of electron transfer between bioreceptor and transducer when used in the electrochemical sensor (Pumera et al. 2011, Kuila et al. 2011, Park et al. 2016, Chauhan et al. 2017). Graphene-based nanomaterials can also generate fluorescent biosensors as they act as quenchers and for this purpose graphene, graphene oxide and reduced graphene oxide are utilized (Kasry et al. 2012, Batir et al. 2018, Wu et al. 2018).

Methods used for the synthesis of graphene-based nanomaterial effect properties and functional role of graphene-based nanomaterial when used in the biosensor. It has also been observed that the number of the graphene sheet layer, functional group presence and oxidation state of graphene-based nanomaterial, and its other derivatives also affect and impact the sensing performance of biosensors and banding in between receptor and transducer. The presence of a functional group on nanomaterial and its amount also affects the detection range of the target molecule. To overcome this limitation, it is of utmost importance to block any nonspecific adsorption site on the nanomaterials by coating it with reagent like bovine serum albumin so that only the target molecule, casein can bind with the nanomaterial (Cheng et al. 2015). Several Graphene nanomaterial-based biosensor has been successfully developed described below.

(i) Antibody Biosensor

Antibody biosensor coupled with graphene-based nanomaterials has been broadly used for the detection of pathogenic microorganisms and used in the biomedical field for the detection of diseases. It was reported that when graphene is modified with other nanoparticles its sensing properties can be enhanced. In this context, graphene is modified by using silver nanoparticles which have been proved useful for the analysis of *Salmonella typhimurium* and viruses like the Hepatitis C virus (Valipour and Roushani 2017). Graphene coupled with gold nanoparticles has been utilized for the detection of influenza virus and oncogenic cells (Dharuman et al. 2013, Huang et al. 2016). Graphene can also be modified with magnetic nanoparticles which have also proved effective for the early detection of diseases like Alzheimer's and Cancer (Demeritte et al. 2015, Sharafeldin et al. 2017).

(ii) Immunosensor

Graphene-based immunosensor has been devised for the detection of microbes in which graphene and graphene oxide is utilized as a sensor platform which reduces the detection limit by ten times.

(iii) DNA Biosensor

Nucleic acid DNA exhibits a wide range of properties like specificity, flexibility, easy synthesis and attachment to the diverse platform, making it suitable to be used for the development of biosensors (Premkumar and Geckeler 2012). The major limitation of DNA to be utilized as a biosensor is the freely degradable nature of DNA which requires specific storage facilities in absence of which its effectiveness can be affected with a minor change in temperature and pH (Koyun et al. 2012). In DNA-based biosensors, graphene-based nanomaterials have been reportedly used as a transducer and two types of DNA-based biosensors, one Electrochemical and another Fluorescent sensor can be developed (Figure 6.6).

(iv) Enzyme-based Biosensor

This type of biosensors are electrochemical in nature and have been devised by enzyme immobilization on graphene surface by using methods, like mixing, sonication, voltammetry and are based on enzymatic properties like catalytic and inhibition/moderation of enzyme activity (Gaudin 2017). Graphene enzyme-based biosensors are utilized to detect a variety of compounds like hydrogen peroxide, glucose, bilirubin and many other compounds. Commonly used enzymes for immobilization are laccase and horse radish peroxidase (Patel et al. 2017) along with other enzymes, like bilirubin oxidase and glucose oxidase can also be immobilized on the surface of graphene.

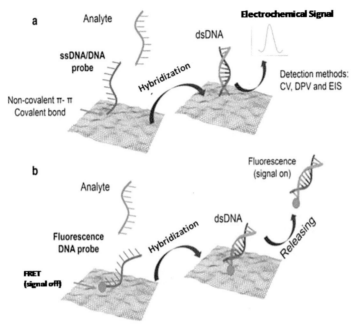

Figure 6.6: Schematic representation of graphene-based nanomaterials uses as DNA biosensors. (a) Electrochemical biosensor (b) and fluorescent biosensor (reproduced with permission from PeñaBahamonde et al. 2018).

3. Conclusion

The application of various nanomaterials for biosensing has revolutionized the field making the technique more user-friendly, cost-effective and smarter by improving detection sensitivity. The unique properties exhibited by nanomaterials offers an alternative to conventional transduction method and ongoing efforts to design more advanced nanomaterial-based biosensor will drastically alter and broaden the field of application of nanomaterial-based biosensor in near future.

References

Afonso, A.S., B. Pérez-López, R.C. Faria, L.H.C. Mattoso, M. Hernández-Herrero, A.X. Roig-Sagués, M.M. da Costa, A. Merkoç. 2013. Electrochemical detection of Salmonella using gold nanoparticles. *Biosens. Bioelectron.* 40, 121–126.

Akter, R., C.K. Rhee, M.A. Rahman. 2015. A highly sensitive quartz crystal microbalance immunosensor based on magnetic bead-supported bienzymes catalyzed mass enhancement strategy. *Biosens. Bioelectron.* 66, 539–546.

Algar, W.R., A.J. Tavares, U.J. Krull. 2010. Beyond labels: A review of the application of quantum dots as integrated components of assays, bioprobes, and biosensors utilizing optical transduction. *Anal. Chim. Acta* 673, 1–25.

Ambrosi, A., M.T. Castañeda, A.J. Killard, M.R. Smyth, S. Alegret, A. Merkoçi. 2007. Double-codified gold nanolabels for enhanced immunoanalysis. *Anal. Chem.* 79, 5232–5240.

Azamian, B.R., J.J. Davis, K.S. Coleman, C.B. Bagshaw, M.L.H. Green. 2002. Bioelectrochemical single-walled carbon nanotubes. *J. of the Amer. Chem. Society* 124(43), 12664–12665.

Bai, L., R. Yuan, Y. Chai, Y. Zhuo, Y. Yuan, Y. Wang. 2012. Simultaneous electrochemical detection of multiple analytes based on dual signal amplification of single-walled carbon nanotubes and multi-labeled graphene sheets. *Biomaterials* 33, 1090–1096.

Bao, D., Z.G. Oh, Z. Chen. 2016. Characterization of silver nanoparticles internalized by Arabidopsis plants using single particle ICP-MS analysis. *Front. Plant. Sci.* 7, 1–8.

Bao, J., X. Geng, C. Hou, Y. Zhao, D. Huo, Y. Wang, Z. Wang, Y. Zeng, M. Yang, H. Fa. 2018. A simple and universal electrochemical assay for sensitive detection of DNA methylation, methyltransferase activity and screening of inhibitors. *J. Electroanal. Chem.* 814, 144–152.

Batır, G.G., M. Arık, Z. Caldıran, A. Turut, S. Aydogan. 2018. Synthesis and characterization of reduced graphene oxide/rhodamine 101 (rGO-Rh101) nanocomposites and their heterojunction performance in rGORh101/p-Si device confguration. *J. Electron. Mater.* 47, 329–336.

Becker, J.S., M. Zoriy, A. Matusch, B. Wu, D. Salber, C. Palm, J.S. Becker. 2010. Bioimaging of metals by laser ablation inductively coupled plasma mass spectrometry (LA-ICP-MS). *Mass Spectrom. Rev.* 29, 156–175.

Beermann, B., E. Carrillo-Nava, A. Scheffer, W. Buscher, A.M. Jawalekar, F. Seela, H.J. Hinza. 2007. Association temperature governs structure and apparent thermodynamics of DNA–gold nanoparticles. *Biophys. Chem.* 126, 124–131.

Brown, K.R., A.P. Fox, M.J. Natan. 1996. Morphology-dependent electrochemistry of cytochrome c at Au colloid-modified SnO2 electrodes. *J. Am. Chem. Soc.* 118, 1154–1157.

Buch, R.M., G.A. Rechnitz. 1989. Intact chemoreceptor-based biosensors: responses and analytical limits. *Biosensors* 4, 215–230.

Buzea, C., I.I. Pacheco, K. Robbie. 2007. Nanomaterials and nanoparticles: Sources and toxicity. *Biointerphases* 2, MR17–MR71.

Cai, H., Y. Wang, P. He, Y. Fang. 2002. Electrochemical detection of DNA hybridization based on silver-enhanced gold nanoparticle label. *Anal. Chim. Acta* 469, 165–172.

Cai, H., Y. Xu, N. Zhu, P. He, Y. Fang. 2002. An electrochemical DNA hybridization detection assay based on a silver nanoparticle label. *Analyst* 127, 803–808.

Cao, C., S.J. Sim. 2007. Signal enhancement of surface plasmon resonance immunoassay using enzyme precipitation-functionalized gold nanoparticles: A femto molar level measurement of anti-glutamic acid decarboxylase antibody. *Biosens. Bioelectron.* 22, 1874–1880.

Chauhan, N., T. Maekawa, D.N.S. Kumar. 2017. Graphene based biosensors accelerating medical diagnostics to new-dimensions. *J. Mater. Res.* 32, 2860–82.

Chen, Y.X., K.J. Huang, F. Lin, L.X. Fang. 2017. Ultrasensitive electrochemical sensing platform based on graphene wrapping SnO2 nanocorals and autonomous cascade DNA duplication strategy. *Talanta* 175, 168–176.

Cheng, S., S. Hideshima, S. Kuroiwa, T. Nakanishi, T. Osaka. 2015. Label-free detection of tumor markers using feldefect transistor (FET) based biosensors for lung cancer diagnosis. *Sens. Actuators B.* 212, 329–34.

Clark, L.C. Jr., R. Wolf, D. Granger, J. Taylor. 1953. Continuous recording of blood oxygen tensions by polarography. *J. Appl. Physiol.* 6(3), 189–193.

Cui, Y., C.M. Lieber. 2001. Functional nanoscale electronic devices assembled using silicon nanowire building blocks. *Science* 291(5505), 851–853.

Cui, D. 2007. Advances and prospects on biomolecules functionalized carbon nanotubes. *J. Nanosci. Nanotechnol.* 7, 1298–1314.

Curie, J., P. Curie. 1880. Development by pressure of polar electricity in hemihedral crystals with inclined faces. *Bull. Soc. Min. Fr.* 3, 90.

Daniel, M.C., D. Astruc. 2004. Gold nanoparticles: assembly, supramolecular chemistry, quantum-size-related properties, and applications toward biology, catalysis, and nanotechnology. *Chem. Rev.* 104, 293–346.

De la Escosura-Muñiz, A., L. Baptista-Pires, L. Serrano, L. Altet, O. Francino, A. Sánchez, A. Merkoci. 2016. Magnetic bead/gold nanoparticle double-labeled primers for electrochemical detection of isothermal amplified leishmania DNA. *Small* 12, 205–213.

Degueldre, C., P.Y. Favarger. 2004. Thorium colloid analysis by single particle inductively coupled plasma-massspectrometry. *Talanta* 62, 1051–1054.

Degueldre, C., P.Y. Favarger, S. Wold. 2006. Gold colloid analysis by inductively coupled plasma-massspectrometry in a single particle mode. *Anal. Chim. Acta* 555, 263–268.

Del Caño, R., L. Mateus, G. Sánchez-Obrero, J.M. Sevilla, R. Madueno, M. Blazquez, T. Pineda. 2018. Hemoglobin becomes electroactive upon interaction with surface-protected Au nanoparticles. *Talanta* 176, 667–673.

Demeritte, T., B.P. Viraka Nellore, R. Kanchanapally, S.S. Sinha, A. Pramanik, S.R. Chavva, P.C. Ray. 2015. Hybrid graphene oxide based plasmonic-magnetic multifunctional nanoplatform for selective separation and label-free identifcation of Alzheimer's disease biomarkers. *ACS Appl. Mater. Interfaces* 7, 13693–700.

Deng, X., M. Chen, Q. Fu, N.M.B. Smeets, F. Xu, Z. Zhang, C.D.M. Filipe, T. Hoare. 2016. A highly sensitive immunosorbent assay based on biotinylated graphene oxide and the quartz crystal microbalance. *ACS Appl. Mater. Interfaces* 8, 1893–1902.

Deng, Z., Y. Zhang, J. Yue, F. Tang, Q. Wei. 2007. Green and orange CdTe quantum dots as effective pH-sensitive fluorescent probes for dual simultaneous and independent detection of viruses. *J. Phys. Chem. B* 111, 12024–12031.

Dharuman, V., J.H. Hahn, K. Jayakumar, W. Teng. 2013. Electrochemically reduced graphene–gold nano particle composite on indium tin oxide for label free immuno sensing of estradiol. *Electrochim Acta* 114, 590–597.

El-Deab, M.S., T. Okajima, T. Ohsaka. 2003. Electrochemical reduction of oxygen on gold nanoparticle electrodeposited glassy carbon electrodes. *J. Electrochem. Soc.* 150, A851–A857.

Fonseca, R.A., J. Ramos-Jesus, L.T. Kubota, R.F. Dutra. 2011. A nanostructured piezoelectric immunosensor for detection of human cardiac troponin T. *Sensors* 11, 10785–10797.

Fu, E., S.A. Ramsey, P. Yager. 2007. Dependence of the signal amplification potential of colloidal gold nanoparticles on resonance wavelength in surface plasmon resonance-based detection. *Anal. Chim. Acta* 599, 118–123.

Gao, Y.S., X.F. Zhu, J.K. Xu, L.M. Lu, W.M. Wang, T.T. Yang, H.K. Xing, Y.F. Yu. 2016. Lu Label-free electrochemical immunosensor based on Nile blue A-reduced graphene oxide nanocomposites for carcino embryonic antigen detection. *Anal. Biochem.* 500, 80–87.

García, M.G., A.C. García. 1995. Adsorptive stripping voltammetric behaviour of colloidal gold and immunogold on carbon paste electrode. *Biosens. Bioelectron.* 38, 389–395.

Gaudin, V. 2017. Advances in biosensor development for the screening of antibiotic residues in food products of animal origin—a comprehensive review. *Biosens. Bioelectronics* 15(90), 363–77.

Geißler, D., L.J. Charbonnière, R.F. Ziessel, N.G. Butlin, H.G. Lchmannsrcben, N. Hildebrandt. 2010. Quantum dot biosensors for ultrasensitive multiplexed diagnostics. *Angew. Chem. Int. Ed.* 49, 1396–1401.

Guiseppi-Elie, A., C. Lei, R.H. Baughman. 2002. Direct electron transfer of glucose oxidase on carbon nanotubes. *Nanotechnology* 13(5), 559–564.

He, L., M.D. Musick, S.R. Nicewarner, F.G. Salinas, S.J. Benkovic, M.J. Natan, C.D. Keating. 2000. Colloidal Au-enhanced surface plasmon resonance for ultrasensitive detection of DNA hybridization. *J. Am. Chem. Soc.* 122, 9071–9077.

He, Q., Z. Zhu, L. Jin, L. Peng, W. Guo, S. Hua. 2014. Detection of HIV-1 p24 antigen using streptavidin–biotin and gold nanoparticles based immunoassay by inductively coupled plasma mass spectrometry. *J. Anal. At. Spectrom.* 29, 1477–1482.

Helfrich, A., W. Brüchert, J. Bettmer. 2006. Size characterisation of Au nanoparticles by ICP-MS coupling techniques. *J. Anal. At. Spectrom.* 21, 431–434.

Heydari-Bafrooei, E., N.S. Shamszadeh. 2017. Electrochemical bioassay development for ultrasensitive apt a sensing of prostate specific antigen. *Biosens. Bioelectron.* 91: 284–292.

Houk, R.S., V.A. Fassel, G.D. Flesch, H.J. Svec, A.L. Gray, C.E. Taylor. 1980. Inductively coupled argon plasma as an ion source for mass spectrometric determination of trace elements. *Anal. Chem.* 52, 2283–2289.

Hu, K., D. Lan, X. Li, S. Zhang. 2008. Electrochemical DNA biosensor based on nanoporous gold electrode and multifunctional encoded DNA-Au bio bar codes. *Anal. Chem.* 80, 9124–9130.

Hu, M., J. Chen, Z.Y. Li, L. Au, G.V. Hartland, X. Li, M. Marquez, Y. Xia. 2006. Gold nanostructures: engineering their plasmonic properties for biomedical applications. *Chem. Soc. Rev.* 35, 1084–1094.

Huang, J., Z. Xie, Z. Xie, S. Luo, L. Xie, L. Huang, Q. Fan, Y. Zhang, S. Wang, T. Zeng. 2016. Silver nanoparticles coated graphene electrochemical sensor for the ultrasensitive analysis of avian infuenza virus H7. *Anal. Chim. Acta* 913, 121–7.

Huang, X., L. Li, H. Qian, C. Dong. J. Ren. 2006. Resonance energy transfer between chemiluminescent donors and luminescent quantum-dots as acceptors (CRET). *Angew. Chem. Int. Ed.* 45, 5140–5143.

Hutter, E., S. Cha, J.F. Liu, J. Park, J. Yi, J.H. Fendler, D. Roy. 2001. Role of substrate metal in gold nanoparticle enhanced surface plasmon resonance imaging. *J. Phys. Chem. B* 105, 8–12.

Ilkhani, H., M. Sarparast, A. Noori, S.Z. Bathaie, M.F. Mousavi. 2015. Electrochemical aptamer/antibody based sandwich immunosensor for the detection of EGFR, a cancer biomarker, using gold nanoparticles as a signaling probe. *Biosens. Bioelectron.* 74, 491–497.

Jackson, S.E., N.J. Pearson, W.L. Griffin, E.A. Belousova. 2004. The application of laser ablation-inductively coupled plasma-mass spectrometry to *in situ* U-Pb zircon geochronology. *Fryer. Chem. Geol.* 211, 47–69.

Janegitz, B.C., T.A. Silva, A. Wong, L. Ribovski, F.C. Vicentini, M.d.P.T. Sotomayor, O. Fatibello-Filho. 2017. The application of graphene for *in vitro* and *in vivo* electrochemical biosensing. *Biosens. Bioelectron.* 89, 224–33.

Jares-Erijman, E.A., T.M. Jovin. 2003. FRET imaging. *Nat. Biotechnol.* 21, 1387–1395.

Jenner, G.A., H.P. Longerich, S.E. Jackson, B.J. Fryer. 1990. ICP-MS—A powerful tool for high-precision trace-element analysis in earth sciences: Evidence from analysis of selected USGS reference samples. *Chem. Geol.* 83, 133–148.

Jeong, B., R. Akter, O.H. Han. 2013. Increased electrocatalyzed performance through dendrimer-encapsulated gold nanoparticles and carbon nanotube-assisted multiple bienzymatic labels: Highly sensitive electrochemical immunosensor for protein detection. *Anal. Chem.* 85, 1784–1791.

Jiang, P., Y. Wang, L. Zhao, C. Ji, D. Chen, L. Nie. 2018. Applications of gold nanoparticles in non-optical biosensors. *Nanomaterials* 8, 977.

Kasry, A., A.A. Ardakani, G.S. Tulevski. 2012. Highly efficient fuorescence quenching with graphene. *J. Phys. Chem. C* 116, 2858–62.

Kim, N., D.K. Kim, Y.J. Cho. 2010. Gold nanoparticle-based signal augmentation of quartz crystal microbalance immunosensor measuring C-reactive protein. *Curr. Appl. Phys.* 10, 1227–1230.

Kim, N.H., T.J. Baek, H.G. Park, G.H. Seong. 2007. Highly sensitive biomolecule detection on a quartz crystal microbalance using gold nanoparticles as signal amplification probes. *Anal. Sci.* 23, 177–181.

Koyun, A., E. Ahlatcolu, Y. Koca. 2012. Biosensors and their Principles. A Road Map Biomed Eng Milestones. New York: InTech.

Kricka, L.J. 1988. Molecular and ionic recognition by biological systems. pp. 3–14. *In*: Edmonds, T.E. (ed.). Chemical Sensors. Blackie and Sons: Glasgow, U.K.

Kuila, T., S. Bose, P. Khanra, A.K. Mishra, N.H. Kim, J.H. Lee. 2011. Recent advances in graphene-based biosensors. *Biosens. Bioelectron.* 26, 4637–4648.

Lau, H.Y., H. Wu, E.J. Wee, M. Trau, Y. Wang, J.R. Botella. 2017. Specific and sensitive isothermal electrochemical biosensor for plant pathogen DNA detection with colloidal gold nanoparticles as probes. *Sci. Rep.* 7, 388–396.

Li, F., J. Han, L. Jiang, Y. Wang, Y. Li, Y. Dong, Q. Wei. 2015. An ultrasensitive sandwich-type electrochemical immunosensor based on signal amplification strategy of gold nanoparticles functionalized magnetic multi-walled carbon nanotubes loaded with lead ions. *Biosens. Bioelectron.* 68, 626–632.

Li, Y., A.W. Wark, H.J. Lee, R.M. Corn. 2006. Single-nucleotide polymorphism genotyping by nanoparticle-enhanced surface plasmon resonance imaging measurements of surface ligation reactions. *Anal. Chem.* 78, 3158–3164.

Li, Y., H.J. Schluesener, S. Xu. 2010. Gold nanoparticle-based biosensors. *Gold Bull.* 43(1), 29–41.

Lin, D., J. Wu, M. Wang. 2012. Triple signal amplification of graphene film, polybead carried gold nanoparticles as tracing tag and silver deposition for ultrasensitive electrochemical immunosensing. *Anal. Chem.* 84, 3662–3668.

Lin, L., Y. Liu, L. Tang, J. Li. 2011. Electrochemical DNA sensor by the assembly of graphene and DNA-conjugated gold nanoparticles with silver enhancement strategy. *Analyst* 136, 4732–4737.

Lin, T.J., K.T. Huang, C.Y. Liu. 2006. Determination of organophosphorous pesticides by a novel biosensor based on localized surface plasmon resonance. *Biosens. Bioelectron.* 22, 513–518.

Link, S. and M.A. El-Sayed. 1996. Spectral properties and relaxation dynamics of surface plasmon electronic oscillations in gold and silver nanodots and nanorods. *J. Phys. Chem. B* 103, 8410–8426.

Liu, B., L. Lu, E. Hua. 2012. Detection of the human prostate-specific antigen using an aptasensor with gold nanoparticles encapsulated by graphitized mesoporous carbon. *Microchim. Acta* 178, 163–170.

Liu, L., N. Xia, H. Liu, X. Kang, X. Liu, C. Xue, X. He. 2014. Highly sensitive and label-free electrochemical detection of microRNAs based on triple signal amplification of multifunctional gold nanoparticles, enzymes and redox-cycling reaction. *Biosens. Bioelectron.* 53, 399–405.

Liu, S.F., J.R. Li, L. Jiang. 2005. Surface modification of platinum quartz crystal microbalance by controlled electroless deposition of gold nanoparticles and its enhancing effect on the HS-DNA immobilization. *Colloid Surf.* 257, 57–62.

Liu, T., J.A. Tang, L. Jiang. 2002. Sensitivity enhancement of DNA sensors by nanogold surface modification. *Biochem. Biophys. Res. Commun.* 295, 14–16.

Liu, T., J.A. Tang, M. Han, L. Jiang. 2003. A novel microgravimetric DNA sensor with high sensitivity. *Biochem. Biophys. Res. Commun.* 304, 98–100.

Liu, T., J.A. Tang, L. Jiang. 2004. The enhancement effect of gold nanoparticles as a surface modifier on DNA sensor sensitivity. *Biochem. Biophys. Res. Commun.* 313, 3–7.

López-Marzo, A.M., R. Hoyos-de-la-Torre, E. Baldrich. 2018. NaNO3/NaCl oxidant and polyethylene glycol (PEG) capped gold nanoparticles (AuNPs) as a novel green route for AuNPs detection in electrochemical biosensors. *Anal. Chem.* 90, 4010–4018.

Lou, Y., M.M. Maye, L. Han, J. Luo, C.J. Zhong. 2001. Gold–platinum alloy nanoparticle assembly as catalyst for methanol electrooxidation. *Chem. Commun.* 0, 473–474.

Lyon, L.A., M.D. Musick, M.J. Natan. 1998. Colloidal Au-enhanced surface plasmon resonance immunosensing. *Anal. Chem.* 70, 5177–5183.

Lyon, L.A., M.D. Musick, P.C. Smith, B.D. Reiss, D.J. Pena, M.J. Natan. 1999. Surface plasmon resonance of colloidal Au-modified gold films. *Sens. Actuators B Chem.* 54, 118–124.

Malik, P., V. Katyal, V. Malik, A. Asatkar, G. Inwati, T.K. Mukherjee. 2013. Nanobiosensors: Concepts and variations. *Nanomaterials* 2013, 1–9.

Matsui, J., K. Akamatsu, N. Hara, D. Miyoshi, H. Nawafune, K. Tamaki, N. Sugimoto. 2005. SPR sensor chip for detection of small molecules using molecularly imprinted polymer with embedded gold nanoparticles. *Anal. Chem.* 77, 4282–4285.

Mcnaught, A.D., A. Wilkinson. 1997. International Union of Pure and Applied Chemistry Compendium of Chemical Terminology; Blackwell Scientific Publications: Oxford, UK, ISBN 0-9678550-9-8.

Miao, X., W. Wang, T. Kang, J. Liu, K.K. Shiu, C.H. Leung, D.L. Ma. 2016. Ultrasensitive electrochemical detection of miRNA-21 by using an iridium(III) complex as catalyst. *Biosens. Bioelectron.* 86, 454–458.

Min, I.H., L. Choi, K.S. Ahn, B.K. Kim, B.Y. Lee, K.S. Kim, H.N. Choi, W.Y. Lee. 2010. Electrochemical determination of carbohydrate-binding proteins using carbohydrate-stabilized gold nanoparticles and silver enhancement. *Biosens. Bioelectron.* 26, 1326–1331.

Morales-Narváez, E., L. Baptista-Pires, A. Zamora-Gálvez, A. Merkoci. 2017. Graphene-based biosensors: going simple. *Adv. Mater.* 29, 1604905.

Mulvaney, P. 1996. Surface plasmon spectroscopy of nanosized metal particles. *Langmuir* 12, 788–800.

Murphy, C.J., A.M. Gole, S.E. Hunyadi, J.W. Stone, P.N. Sisco, A. Alkilany, B.E. Kinard, P. Hankins. 2008. Chemical sensing and imaging with metallic nanorods. *Chem. Commun. (Camb.)* 5, 544.

Murray, C.B., D.J. Norris, M.G. Bawendi. 1993. Synthesis and characterization of nearly monodisperse CdE (E = sulfur, selenium, tellurium) semiconductor nanocrystallites. *J. Am. Chem. Soc.* 115, 8706–8715.

Ng, B.Y., W. Xiao, N.P. West, E.J.H. Wee, Y. Wang, M. Trau. 2015. Rapid, single-cell electrochemical detection of Mycobacterium tuberculosis using colloidal gold nanoparticles. *Anal. Chem.* 87, 10613–10618.

Ozsoz, M., A. Erdem, K. Kerman, D. Ozkan, B. Tugrul, N. Topcuoglu, H. Ekren, M. Taylan. 2003. Electrochemical genosensor based on colloidal gold nanoparticles for the detection of Factor V Leiden mutation using disposable pencil graphite electrodes. *Anal. Chem.* 75, 2181–2187.

Pan, B., D. Cui, C.S. Ozkan, M. Ozkan, P. Xu, T. Huang, F. Liu, H. Chen, Q. Li, R. He, F. Gao. 2008. Effects of carbon nanotubes on photoluminescence properties of quantum dots. *J. Phys. Chem. C* 112, 939–944.

Pan, Y., W. Shan, H. Fang, M. Guo, Z. Nie, Y. Huang, S. Yao. 2014. Sensitive and visible detection of apoptotic cells on Annexin-V modified substrate using aminophenylboronic acid modified gold nanoparticles (APBA-GNPs) labeling. *Biosens. Bioelectron.* 52, 62–68.

Pandit, S., D. Dasgupta, N. Dewan, P. Ahmed. 2016. Nanotechnology based biosensors and its application. *The Pharma Inno. Journal* 5(6), 18–25.

Park, C.S., H. Yoon, O.S. Kwon. 2016. Graphene-based nanoelectronic biosensors. *J. Ind. Eng. Chem.* 38, 13–22.

Park, J., J. Joo, S.G. Kwon, Y. Jang, T. Hyeon. 2007. Synthesis of monodisperse spherical nanocrystals. *Angew. Chem. Int. Ed.* 46, 4630–4660.

Patel, S.K., S.H. Choi, Y.C. Kang, J.K. Lee. 2017. Eco-friendly composite of Fe3O4-reduced graphene oxide particles for efficient enzyme immobilization. *ACS Appl. Mater. Interfaces* 9: 2213–2222.

PeñaBahamonde, J., H.N. Nguyen, S.K. Fanourakis, D.F. Rodrigues. 2018. Recent advances in graphene based biosensor technology with applications in life sciences. *J. Nanobiotechnol.* 16, 75.

Premkumar, T., K.E. Geckeler. 2012. Graphene–DNA hybrid materials: assembly, applications, and prospects. *Prog. Polym. Sci.* 37, 515–29.

Pumera, M., M.T. Castaneda, M.I. Pividori, R. Eritja, A. Merkoci, S. Alegret. 2005. Magnetically trigged direct electrochemical detection of DNA hybridization using Au67 quantum dot as electrical tracer. *Langmuir* 21, 9625–9629.

Pumera, M. 2011. Graphene in biosensing. *Mater. Today* 14, 308–315.

Qin, X.A.Xu., L. Liu, W. Deng, C. Chen, Y. Tan, Y. Fu, Q. Xie, S. Yao. 2015. Ultrasensitive electrochemical immunoassay of proteins based on *in situ* duple amplification of gold nanoparticle biolabel signals. *Chem. Commun.* 51, 8540–8543.

Qu, H., L. Yang, J. Yu, L. Wang, H. Liu. 2018. Host-guest interaction induced rapid self-assembled Fe3O4@Au nanoparticles with high catalytic activity. *Ind. Eng. Chem. Res.* 57, 9448–9456.

Rao, H., Y. Liu, J. Zhong, Z. Zhang, X. Zhao, X. Liu, Y. Jiang, P. Zou, X. Wang, Y. Wang. 2017. Gold nanoparticle/chitosan@N,S co-doped multiwalled carbon nanotubes sensor: Fabrication, characterization, and electrochemical detection of catechol and nitrite. *ACS Sustain. Chem. Eng.* 5, 10926–10939.

Schierhorn, M., S.J. Lee, S.W. Boettcher, G.D. Stucky, M. Moskovits. 2006. Metal Silica hybrid nanostructures for surface enhanced raman spectroscopy. *Adv. Mater.* 18, 28–29.

Sharafeldin, M., G.W. Bishop, S. Bhakta, A. El-Sawy, S.L. Suib, J.F. Rusling. 2017. Fe3O4 nanoparticles on graphene oxide sheets for isolation and ultrasensitive amperometric detection of cancer biomarker proteins. *Biosens. Bioelectron.* 91, 359–366.

Shi, L., X. Rong, Y. Wang, S. Ding, W. Tang. 2018. High-performance and versatile electrochemical aptasensor based on self-supported nanoporous gold microelectrode and enzyme-induced signal amplification. *Biosens. Bioelectron.* 102, 41–48.

Shu, H., W. Wen, H. Xiong, X. Zhang, S. Wang. 2013. Novel electrochemical aptamer biosensor based on gold nanoparticles signal amplification for the detection of carcinoembryonic antigen. *Electrochem. Commun.* 37, 15–19.

Shuai, H.L., K.J. Huang, L.L. Xing, Y.X. Chen. 2016. Ultrasensitive electrochemical sensing platform for microRNA based on tungsten oxide-graphene composites coupling with catalyzed hairpin assembly target recycling and enzyme signal amplification. *Biosens. Bioelectron.* 86, 337–345.

Song, W., X. Guo, W. Sun, W. Yin, P. He, X. Yang, X. Zhang. 2018. Target-triggering multiple-cycle signal amplification strategy for ultrasensitive detection of DNA based on QCM and SPR. *Anal. Biochem.* 553, 57–61.

Takae, S., Y. Akiyama, Y. Yamasaki, Y. Nagasaki, K. Kataoka. 2007. Colloidal Au replacement assay for highly sensitive quantification of low molecular weight analytes by surface plasmon resonance. *Bioconjug. Chem.* 18, 1241–1245.

Tang, D., B. Zhang, J. Tang, L. Hou, G. Chen. 2013. Displacement-type quartz crystal microbalance immunosensing platform for ultrasensitive monitoring of small molecular toxins. *Anal. Chem.* 85, 6958–6966.

Taylor, R., C. Sylvain, O. Todd, S. Coulombe, T. Otanicar, P. Phelan, A. Gunawan, W. Lv, G. Rosengarten, R. Prasher, H. Tyagi. 2013. Small particle, big impacts: a review of the diverse applications of nanofluids. *J. Appl. Phys.* 113(1), 011301.

Tseng, Y.T., Y.J. Chuang, Y.C. Wu. 2008. Biosensor based on selected gold nanoparticles. *Nanotechnology* 19, 345–501.

Turner, A.P.F., I. Karube, G.S. Wilson. 1987. Biosensors—Fundamentals and Applications, Oxford University Press: New York, NY, USA, 719–800.

Umar, A., M.M. Rahman, A. Al-Hajry, Y.B. Hahn. 2009. Highly sensitive cholesterol biosensor based on well-crystallized flower-shaped ZnO nanostructures. *Talanta* 78(1), 284–289.

Valden, M., X. Lai, D.W. Goodman. 1998. Onset of catalytic activity of gold clusters on titania with the appearance of nonmetallic properties. *Science* 281, 1647–1650.

Valentini, F., M. Carbone, G. Palleschi. 2013. Carbon nanostructured materials for applications in nano-medicine, cultural heritage, and electrochemical biosensors. *Anal. Bioanal. Chem.* 405, 451–465.

Valipour, A., M. Roushani. 2017. Using silver nanoparticle and thiol graphene quantum dots nanocomposite as a substratum to load antibody for detection of hepatitis C virus core antigen: electrochemical oxidation of ribofavin was used as redox probe. *Biosens. Bioelectron.* 89, 946–51.

Vural, T., Y.T. Yaman, S. Ozturk, S. Abaci, E.B. Denkbas. 2018. Electrochemical immunoassay for detection of prostate specific antigen based on peptide nanotube-gold nanoparticle-polyaniline immobilized pencil graphite electrode. *J. Colloid Interface Sci.* 510, 318–326.

Wang, J., R. Polsky, D. Xu. 2001. Silver-enhanced colloidal gold electrochemical stripping detection of DNA hybridization. *Langmuir* 17, 5739–5741.

Wang, J. 2002. Electrochemical nucleic acid biosensors. *Anal. Chim. Acta* 469(1), 63–71.

Wang, J., D. Xu, R. Polsky. 2002. Magnetically-induced solid-state electrochemical detection of DNA hybridization. *J. Am. Chem. Soc.* 124, 4208–4209.

Wang, J., J. Li, A.J. Baca, J. Hu, F. Zhou, W. Yan, D.W. Pang. 2003. Amplified voltammetric detection of DNA hybridization via oxidation of ferrocene caps on gold nanoparticle/streptavidin conjugates. *Anal. Chem.* 75, 3941–3945.

Wang, J., K. Ma, H. Yin, Y. Zhou, S. Ai. 2018. Aptamer based voltammetric determination of ampicillin using a single-stranded DNA binding protein and DNA functionalized gold nanoparticles. *Microchim. Acta* 185, 68.

Wang, L., Q. Wei, C. Wu, Z.Y. Hu, J. Ji, P. Wang. 2008. The Escherichia coli O157: H7 DNA detection on a gold nanoparticle-enhanced piezoelectric biosensor. *Chin. Sci. Bull.* 53, 1175–1184.

Wang, L., E. Hua, M. Liang, C. Ma, Z. Liu, S. Sheng, M. Liu, G. Xie, W. Feng. 2014. Graphene sheets, polyaniline and AuNPs based DNA sensor for electrochemical determination of BCR/ABL fusion gene with functional hairpin probe. *Biosens. Bioelectron.* 51, 201–207.

Wang, X., D. Du, H. Dong, S. Song, K. Koh, H. Chen. 2018. para-Sulfonatocalix[4]arene stabilized gold nanoparticles multilayers interfaced to electrodes through host-guest interaction for sensitive ErbB2 detection. *Biosens. Bioelectron.* 99, 375–381.

Wang, Y., E.C. Alocilja. 2015. Gold nanoparticle-labeled biosensor for rapid and sensitive detection of bacterial pathogens. *J. Biol. Eng.* 9, 16.

Wang, Z., J. Zhang, C. Zhu, S. Wu, D. Mandler, R.S. Marks, H. Zhang. 2014. Amplified detection of femtomolar DNA based on a one-to-few recognition reaction between DNA-Au conjugate and target DNA. *Nanoscale* 6, 3110–3115.

Wang, W., C. Ma, Y. Li, B. Liu, L. Tan. 2018. One-pot preparation of conducting composite containing abundant amino groups on electrode surface for electrochemical detection of von willebrand factor. *Appl. Surf. Sci.* 433: 847–854.

Weller, H. 1993. Colloidal semiconductor q-particles: chemistry in the transition region between solid state and molecules. *Angew. Chem. Int. Ed. Engl.* 32, 41–53.

Wu, X., Y. Xing, K. Zeng, K. Huber, J.X. Zhao. 2018. Study of fuorescence quenching ability of graphene oxide with a layer of rigid and tunable silica spacer. *Langmuir* 34, 603–11.

Xiao, Y., F. Patolsky, E. Katz, J.F. Hainfeld, I. Willner. 2003. Plugging into enzymes: Nanowiring of redox enzymes by a gold nanoparticle. *Science* 299, 1877.

Xu, G., H. Li, X. Ma, X. Jia, J. Dong, W. Qian. 2009. A cuttlebone-derived matrix substrate for hydrogen peroxide/glucose detection. *Biosens. Bioelectron.* 25, 362–367.

Yan, Z., M. Yang, Z. Wang, F. Zhang, J. Xia, G. Shi, L. Xia, Y. Li, Y. Xia, L. Xia. 2015. A label-free immunosensor for detecting common acute lymphoblastic leukemia antigen (CD10) based on gold nanoparticles by quartz crystal microbalance. *Sens. Actuators B Chem.* 210, 248–253.

Yang, W., Z. Xi, X. Zeng, L. Fang, W. Jiang, Y. Wu, L.J. Xua, F.F. Fu. 2016. Magnetic bead-based AuNP labelling combined with inductively coupled plasma mass spectrometry for sensitively and specifically counting cancer cells. *J. Anal. At. Spectrom.* 31, 679–685.

Yang, X., Q. Wang, K. Wang, W. Tan, H. Li. 2007. Enhanced surface plasmon resonance with the modified catalytic growth of Au nanoparticles. *Biosens. Bioelectron.* 22, 1106–1112.

Yin, H., Y. Zhou, H. Zhang, X. Meng, S. Ai. 2012. Electrochemical determination of microRNA-21 based on graphene, LNA integrated molecular beacon, AuNPs and biotin multifunctional bio bar codes and enzymatic assay system. *Biosens. Bioelectron.* 33, 247–253.

Yuan, H., S. Gao, X. Liu, H. Li, D. Günther, F. Wu. 2004. Accurate U-Pb age and trace element determinations of zircon by laser ablation-inductively coupled plasma-mass spectrometry. *Geostand. Geoanal. Res.* 28, 353–370.

Zhang, S., H. Bai, J. Luo, P. Yang, J. Cai. 2014. A recyclable chitosan-based QCM biosensor for sensitive and selective detection of breast cancer cells in real time. *Analyst* 139, 6259–6265.

Zhang, X., J. Shen, H. Ma, Y. Jiang, C. Huang, E. Han, B. Yao, Y. He. 2016. Optimized dendrimer-encapsulated gold nanoparticles and enhanced carbon nanotube nanoprobes for amplified electrochemical immunoassay of *E. coli* in dairy product based on enzymatically induced deposition of polyaniline. *Biosens. Bioelectron.* 80, 666–673.

Zhang, Y., M. Yang, N.G. Portney, D. Cui, G. Budak, E. Ozbay, M. Ozkan, C.S. Ozkan. 2008. Zeta potential: A surface electrical characteristic to probe the interaction of nanoparticles with normal and cancer human breast epithelial cells. *Biomed. Microdev.* 10, 321–328.

Zhang, Y., J. Xiao, Y. Sun, L. Wang, X. Donga, J. Ren, W. He, F. Xiao. 2018. Flexible nanohybrid microelectrode based on carbon fiber wrapped by gold nanoparticles decorated nitrogen doped carbon nanotube arrays: *In situ* electrochemical detection in live cancer cells. *Biosens. Bioelectron.* 100, 453–461.

Zhao, H., L. Lin, T. Ji'an, D. Mingxing, J. Long. 2001. Enhancement of the immobilization and discrimination of DNA probe on a biosensor using gold nanoparticles. *Chin. Sci. Bull.* 46, 1074–1077.

Zhao, H.Q., L. Lin, J.R. Li, J.A. Tang, M.X. Duan, L. Jiang. 2001. DNA biosensor with high sensitivity amplified by gold nanoparticles. *J. Nanopart. Res.* 3, 321–323.

Zhao, J., Y. Zhang, H. Li, Y. Wen, X. Fan, F. Lin, L. Tan, S. Yao. 2011. Ultrasensitive electrochemical aptasensor for thrombin based on the amplification of aptamer–AuNPs–HRP conjugates. *Biosens. Bioelectron.* 26, 2297–2303.

Zhao, L., Z. Ma. 2017. New immunoprobes based on bovine serum albumin-stabilized copper nanoclusters with triple signal amplification for ultrasensitive electrochemical immunosensing for tumor marker. *Sens. Actuators B Chem.* 241, 849–854.

Zhao, Y., H. Wang, W. Tang, S. Hu, N. Lia, F. Liu. 2015. An *in situ* assembly of a DNA–streptavidin dendrimer nanostructure: A new amplified quartz crystal microbalance platform for nucleic acid sensing. *Chem. Commun.* 51, 10660–10663.

Zheng, J., W. Feng, L. Lin, F. Zhang, G. Cheng, P. He, Y. Fang. 2007. A new amplification strategy for ultrasensitive electrochemical aptasensor with network-like thiocyanuric acid/gold nanoparticles. *Biosens. Bioelectron.* 23, 341–347.

Zheng, X., L. Li, K. Cui. 2018. Ultrasensitive enzyme-free biosensor by coupling cyclodextrin functionalized Au nanoparticles and high-performance Au-paper electrode. *ACS Appl. Mater. Interfaces* 10, 3333–3340.

Zhou, Y., Q. Xie. 2016. Hyaluronic acid-coated magnetic nanoparticles-based selective collection and detection of leukemia cells with quartz crystal microbalance. *Sens. Actuators B Chem.* 223, 9–14.

Zhou, Y., H. Yin, X. Li, Z. Li, S. Ai, H. Lin. 2016. Electrochemical biosensor for protein kinase A activity assay based on gold nanoparticles-carbon nanospheres, phos-tag-biotin and β-galactosidase. *Biosens. Bioelectron.* 86, 508–515.

Zong, Y., F. Liu, Y. Zhang, T. Zhan, Y. He, X. Hun. 2016. Signal amplification technology based on entropy-driven molecular switch for ultrasensitive electrochemical determination of DNA and Salmonella typhimurium. *Sens. Actuators B Chem.* 225, 420–427.

Chapter **7**

Nanomaterials Application in Biological Sensing of Biothreat Agents

1. Introduction

In the high-tech world, we are living today human population is threatened by unceasing exposure to diseases either induced by pathogenic bioagents premeditated or by the sudden outbreak of diseases which are mortal. Prompt detection and meticulous diagnosis at the onset of the disease is the most expedient approach to save the life of a person (Zhou et al. 2015, Tripathi et al. 2016, Susana et al. 2017) and for this, brisk economical and reliable diagnostic methods are recommended. Among the different microscopic techniques developed in recent years, biosensors are one of them.

The term biosensor was first used by Professor Karl Camann in 1977, but its function was not specified until 1997 when the International Union of Pure and Applied Chemistry (IUPAC) delineated biosensor signifying its role as a sensitive analytical device which can give decisive diagnosis result (Islam and Uddin 2017). Biosensor coalesced with nanomaterials are applied in the field of diagnostic in medical science favorably.

(a) Nonmaterial-Based Biosensors For Detection of Bioagents

For a long time, microorganisms are known to have caused many life-threatening diseases in humans and a sudden outbreak of these diseases in the history of mankind has claimed the life of masses. The breakthrough in research in the area of microbiology has propelled extensive research to develop microorganisms as biothreat agents and their production as bioweapons. In 1972, the United Nations organized a Biological Weapon Convention (BWC) in which production and development of Bioweapon were rigorously banned (Dembek 2007, Georgiev 2009, Dembek 2011, Jin and Hildebrandt 2012). Since then there are many reports about the use of bioagents in the act of bioterrorism enlisted in Table 7.1.

Bioagents contain any of the pathogenic microorganisms like Bacteria, Viruses and Toxins which are noxious for humans. The conventional detection method which recognizes and analyzes the presence of biological agents in a sample includes biochemical, immunological assay and polymerase chain reaction (PCR), but early detection of these biological agents is of prime importance to save the human population from threat in which these conventional technique dearths. So in recent years, much efforts have been given to diagnostic technique which can give a rapid and meticulous result within a limited period by exploiting exceptional properties of miniaturized nanomaterials and a range of biosensors based on nanomaterials have been developed.

2. What are Bioagents?

Bioagents or biological agents are pathogens microorganism or their toxic products which are a menace for human health (LeClaire and Pitt 2005, Dembek 2007, Sapsford et al. 2008, Dembek 2011). They include both naturally occurring and man-made bioengineered pathogens that induce health risk from their natural outbreak or premeditated release. These agents are deliberately used in biowarfare or bioterrorism due to an array of peculiar features like:

➤ Infection efficiency signifies how easily it can establish itself with the host.

➤ Virulence signifies the severity of the disease caused by the microorganism.

➤ The contagious nature of a pathogen signifies how deliberately it can get disseminated from one person to another causing the infection (Dembek 2007, Dembek 2011).

Other important aspects are pathogenicity, inoculation period, etc., which make a microorganism suited as a bioweapon.

3. Some Common Bioagents as Biothreat

Among the vast diversity of microorganisms, there are some which are highly pathogenic and cause severe diseases in human.

Table 7.1: List of biothreat agents (Walper et al. 2018, CDC's Chemical Emergencies/Agents list https://emergency.cdc.gov/agent/agentlist-category.asp).

Pathogen	Disease Caused	CDC Bioterrorism/Agent List (Yes/No)
Bacterial Agents		
Bacillus anthracis	Anthrax	Yes
Brucella spp.	Brucellosis	Yes
Burkholderia spp.		
• *B. mallei*	Glanders	Yes
• *B. pseudomallei*	Melioidosis	Yes
Chlamydia psittaci	Psittacosis	Yes
Clostridium botulinum	Botulism	Yes
Coxiella burnetii	Q-fever	Yes
Francisella tularensis	Tularemia	Yes
Mycobacterium tuberculosis	Tuberculosis, TB	No
Rickettsia prowazeki	Typhus fever	Yes
Yersinia pestis	Plague	Yes
Viral Agents		
Chikungunya virus	Chikungunya	No
Coronavirus (CoV)	SARS, MERS, COVID-19	No
Influenza viruses	Influenza	No
Encephalitis viruses	Encephalitis	Yes
Rabies virus	Rabies	No
Arena viruses	Viral hemorrhagic fever	Yes
Yellow fever virus	Yellow fever	No
Toxin Agents		
Abrin		No
Neurotoxins		Yes
Conotoxins		No
Epsilon toxin		Yes
Ricin		Yes
Saxitoxin		No
Staphylococcal enterotoxins		Yes
Tetrodotoxin		No
T-2 toxin		No

(a) Bacteria

Bacteria are small unicellular microorganisms that can replicate and are a casual agent of many diseases. Their use as bioweapon started during World War I when animals shipped from the U.S to allied forces were infected by Anthrax powder and Glanders by the German agents (Riedel 2004). Bacteria cause a wide range of diseases with a high fatality rate.

(i) Some Common Bacterial Biothreats

➢ **Anthrax**—The first report of a natural outbreak of anthrax dates back to the 1st century and was reported in 700 BC by Homer in Iliad and Virgil (www.cdc.gov 2018). From 1978 through 1980, an epidemic was reported in Zimbabwe due to anthrax infection acquired from infected animals and reportedly almost 10,000 people were infected (Dembek 2007, Dembek 2011).

From that time onwards, much research has been conducted on *Bacillus anthracis* by countries like the U.S, the U.K and then USSR (Figure 7.1). Along with other toxins, *B. anthracis* was cultured by Iraqi forces led by Saddam Hussein as a bioweapon and were armed into bombs and missiles used during the Persian Gulf War (Dembek 2007). A recent act of bioterrorism in 2011 was reported from the United States where spores of *B. anthracis* were used to infect the letter mailed to U.S. congress and press infecting around 22 people in contact. Anthrax spores exhibit an incubation period of 1–7 days (NATO 1996), and medication given to the infected person during the incubation period can be effective but once after the onset of symptoms antibiotics do not show impressive results (NATO 1996, LeClaire and Pitt 2005).

➢ **Plague**—In the history of mankind natural incidence of plague spontaneous outbreak were culpable for three severe pandemics, including the infamous Black Death which ended the life of masses (Figure 7.1). A notable use of plague as bioweapon was reported during World War II and was used by Japanese troops. Japan biowarfare Unit 731 experimented with *Yersinia pestis* and dispersed it during the period of war through fleas over Chinese cities which culminated in Bubonic plague with a fatality rate of 5% (Georgiev 2009). An alternative plague, i.e., pneumonic plague has a high fatality rate of around 50% and has proved other potent bioagent (Bearden et al. 1997, Chu 2001).

➢ **Glanders/Melioidosis**—This disease is caused by bacteria *Burkholderia mallei* found in horses but can also cause grievous sickness in humans if transmitted. Transmission cases from horses to human is occasional,

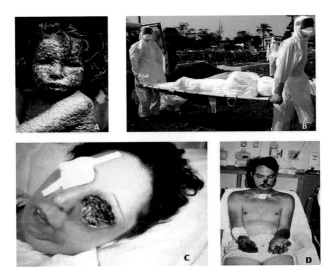

Figure 7.1: Images of patients infected with biothreat agents. (A) A young girl was infected with smallpox in 1973. (B) An illustrative image from a cemetery in Sierra Leone's, representing the burial process with appropriate measures during the Ebola outbreak to prevent further spreading of the Virus. (C) A patient suffering from Anthrax infection of the eye. (D) A patient infected with Yersinia pestis caused the black coloration of the body. (Reproduced with permission from Walper et al. 2018).

and the last incidence was recorded in 1934. They are still high-risk biothreat agents due to their infection efficiency and accessibility (Wittig et al. 2006, Khan et al. 2013).

➤ **Tularaemia**—It is also a bacterial disease caused by *Francisella tularensis* and infects both animals (Rodents and Rabbits) and humans. This bacteria causes severe malady like prognosis which has a very high fatality rate, thus making it a potent bioagent for biothreat. Naturally occurring incidence of tularaemia was recorded from the United States (Hestvik 2015). According to some reports, these bioagents were also used as bioweapon during World War II and the Japanese biothreat fare unit also conducted several experiments on these bioagents (Croddy et al. 2001, Reintjes 2002, Hestvik 2015).

➤ **Q Fever**—Another bacterial disease caused by *Coxiella burnetii* was initially discovered in countries like Australia and the U.S. before World War II. Primarily zoonotic in origin, they are transmitted from animals like cows and sheep to humans (Baca and Paretsky 1983, deRooij et al. 2016).

➤ **Bacteria as threats to food and water**—Many diseases causing bacteria to infiltrate in our food and water supply system and cause several diseases. Some common bacteria are *Clostridium botulinum, E. coli, Salmonella typhi, Vibrio cholera* and *Campylobacter jejuni*. The detection of these bacteria is based on the recognition of toxins produced by these bacteria.

(b) Viruses as Biothreat Agents

Viruses are deemed as highly infectious bioagents comprising DNA or RNA as their genetic material enclosed within a shell of protein. Viruses commonly require a host for multiplication (Cheng et al. 2009, Condit 2001). By virtue of their highly infectious nature, high fatality rate caused by viral diseases and transmission competency from person to person make the man impressive candidate for bioagents to be used as bioweapons.

An endemic outbreak due to a virus follows a seasonal trend. The year 1918–19 witnessed the outbreak of Spanish flu causing around 20–50 million deaths (Kaiser 2005) and the year 2019–20 has also witnessed a pandemic situation due to virus outbreak.

(i) Some Common Viral Bioagents

➢ **Smallpox**—The disease caused by *Variola* virus was first identified and described by Duggan and his research group while he was working on the remains of a seventeenth century mummy of a child (Duggan et al. 2016). Smallpox is an eminently contagious disease and on average kills three people out of each ten affected. It has been reported by several researchers that smallpox has claimed the lives of nearly 500 million people in the twentieth century (Dembek 2007). Considering all evidence WHO initiated the smallpox eradication program in 1966 to make the world free from smallpox which was successful.

➢ **Viral Hemorrhagic Fever**—This disease is caused by single-stranded RNA viruses belonging to four distinct families. Among them, Ebola and Marburg are well known hemorrhagic virus belonging to the family Filoviridae on account of their filament-like body (Dembek 2007, 2011). Although the explicit natural reservoir of these viruses is obscure, bats have been considered and reported as the most likely source (Morvan et al. 1999, Leroy et al. 2005, Towner et al. 2009). These bioagents are considered significant threats when used in the form of bioweapons due to their high fatality rate (Feldmann 2014).

➢ **Influenza**—The virus causing influenza belongs to family Orthomyxoviridae which consist of three genera A, B and C. These three genera in consolidation are liable for causing influenza in human. Reports suggest that genera A and C can infect birds as well as mammals (Knipe and Howley 2001). As they are RNA viruses, they can mutate quickly which has driven evolution in the strains which in recent years are responsible for the appearance of H5N1 and H7N9 (Avian) flue and H1N1 and N3N2 (Swine) flue. To date, the use of influenza viruses as bioagent in biowarfare or bioterrorism cases has not been reported, but its spontaneous outbreak claims the life of 25,000–500,000 people every year.

> **Coronavirus and Acute Respiratory Syndrome**—The first notable spontaneous outbreak of SARS was reported in 2003 from Asia. During the first year of the outbreak, the fatality rate was 10% (Dostalova 2015). Another spontaneous outbreak of coronavirus resulted in infection-causing MERS disease which was reported in 2012 from Saudi Arabia (WHO 2015). Currently, the case of coronavirus outbreak was reported from China at the end of December 2019 causing infection Covid-19 which has claimed the lives of many people so far across the globe. SARS-CoV, MERS-CoV and SARS-CoV belong to the family Coronaviridae and are enveloped by protective protein covering that comprise of single-stranded RNA as their genetic material (King et al. 2012, Hilgenfeld and Peiris 2013). The natural source of these viruses is bats or civets (de Groot et al. 2013).

(c) Toxins as Bioagents

Toxins are deadly substances produced by biological sources which include animals, plants and microorganisms. Toxins are found both in proteinaceous or chemical form (Walper et al. 2018). As they lack the power of self-replication so toxin-based biothreats are usually based on their dosage percentage, fatality effect and mode of delivery. Extraction of toxins requires efficient production techniques and depository facilities as some toxins get denatured easily by environmental factors like heat and radiation as reported.

(i) Some Common Toxins as Biothreat Agents

> **Botulinum Neurotoxin (BoNT)**—At the current time, BoNT is among the three widely researched biothreat agents. For use in form of bioweapon, Aum Shinrikyo from Japan cultured the bacteria *Clostridium botulinum* producing BoNT and was successful in his research. He also did several experiments with the dispersion system (Dembek 2007). The toxin when administered in the human body a neuroparalytic disease called Botulism develops. The toxin is extremely lethal and the fatality percentage is very also very high.

➢ **Ricin**—It is a proteinaceous substance that is highly toxic and is derived from the seeds of *Ricinus communis*. It is less potent as compared to BoNT, but the worldwide availability of seeds of *Ricinus* makes it a promising candidate as a biothreat agent. Ricin can be delivered either orally, by inhalation or through intramuscular routes, and its show different effect depending on the route of administration (Schep et al. 2009, Harkup 2015).

➢ **Abrin**—It is a ribosome-inactivating toxic protein extracted from plant *Abrus precatorius* (Lin et al. 1970). It is also the most probable candidate for use as a biothreat agent having toxic characterization the same as Ricin, but at the same time, it is 75% more toxic than Ricin (Patocka 2001).

➢ **Staphylococcal Enterotoxin B**—The most common source of food poising produced by the bacteria *Staphylococcus aureus*. It severely affects the immune system and even a small dosage is efficient in producing fatal effects.

➢ **Mycotoxins**—These are a class of toxin which are produced by Fungi. A mycotoxin is a low molecular weight proteinaceous compound and its main source is filamentous molds. The use of mycotoxin as a biothreat agent was reported during the Cold War in 1975 and 1981 causing almost 10,000 suspected deaths. It is generally disseminated through the nasal route (St. Georgiev 2009) causing some severe symptoms like inflammation of skin and eyes, the toxicity of the central nervous system (NATO 1996, Dembek 2007).

➢ **Tetrodotoxin**—This toxin is extracted from Pufferfish which is an important part of Japanese cuisine. Pufferfish produces this toxin in the symbiotic presence of some bacteria. If its amount is high (2 mg or more), it results in death (Moczydlowski 2013). Its use as bioweapon could be in aerosol form.

4. Nanomaterial-Based Biosensor for Detection of Bioagents

The current detection method which is available like ELISA, PCR, Immunoassay, etc., require a long duration for a complete analysis, but the need for the hour is the development of better sensing techniques with improved sensitivity, which should be less time consuming and also very cost-effective. Nanomaterials due to their small size have proven to be a sterling candidate for use in biosensors as they provide large surface area and give the result of analysis in less than a minute (Vaseashta et al. 2012). Nanomaterials also possess unique photophysical features that can be exploited for tagging of analyte and transfer of energy. Many nanomaterials have been used in biosensors and discussed further.

(a) Nanomaterial Probe

Nanomaterials are biocompatible with a wide variety of biomolecules and can be effectively utilized as a probe for the detection of biothreat agents. The biofunctionalization property of nanomaterials allows recognition and attachment of biomolecules on their surface through electrostatic, covalent and non-covalent bonding (Field et al. 2015).

These properties permit in designing probes that are proficient in detecting a wide variety of bioagents. Bacteria can be easily detected through antibodies binding to different moieties, including protein, carbohydrates and sugar displayed on the cell surface. Oligonucleotide coated on the surface of nanoparticles is also used to apprehend the genetic material of bioagent which is later amplified by PCR (Medintz 2006).

(b) Gold Nanoparticles (AuNPs)

Gold nanoparticles are extensively used for the detection of bioagents like bacteria, viruses and toxic products (Upadhyayula 2012). AuNPs are found compatible with a vast variety of biological and chemical molecules. AuNPs of size ranging between 2–100 nm exhibit a special phenomenon that is known as Localized Surface Plasmon Resonance because of which AuNPs show exemplary optical absorption properties, thus enhancing the sensing ability of AuNPs. Gold nanoparticles of size below 2 mm are used to differentiate between materials exhibiting fluorescent and plasmonic signals.

AuNP which are used for detection can be functionalized with an oligonucleotide that is complementary to the target DNA. It was demonstrated by Sattarahmady and his group that detection of *Brucella* species was complete within a time limit of 10 minutes in contrast to hours taken by PCR or cell culture technique (Figure 7.2). In the presence of the analyte, AuNP forms a cluster giving rise to some calorimetric change which cannot be noticed in the presence of any non-complementary DNA (Sattarahmady et al. 2015). As the change in color is apparent, using absorption spectroscopy can provide valuable analytical results. This scheme can be easily converted into a detection system for other bacterial species just by changing surface functionalization and antibody use (Guarise et al. 2006, Gill et al. 2008, Thiruppathiraja et al. 2011, Hussain et al. 2013). Aggregation of AuNPs in response to electrochemical change has been used for the detection of varieties of other bioagents (Owino

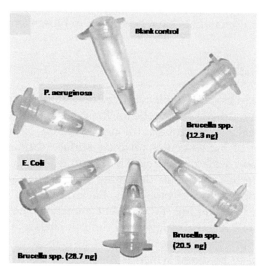

Figure 7.2: Gold nanoparticles aggregate in presence of Brucella spp. Any visual change in the color of the aggregate is not noticed exposed to non-complementary DNA from any other bacterial species (Reproduced with permission from Sattarahmady et al. 2015).

et al. 2008, Thiruppathiraja et al. 2011, Upadhyayula 2012, Saha et al. 2012, Jarocka et al. 2014).

AuNPs loaded with antibodies for the detection of toxin ricin works well as reported in the case of a sandwich immunoassay with output in the form of bands (Shyu et al. 2002). Surface embellished Raman Spectroscopy used in combination with AuNPs is also used for the detection of bioagents (Cao et al. 2002, Stoeva et al. 2006, Upadhyayula 2012, Saha et al. 2012). Silver present on AuNP surface improves SERS (Surface-enhanced Raman Scattering) signal, thus lowering the detection limit (Cao et al. 2002). Another property of nanomaterials, i.e., surface plasmon resonance (SPR) is also used in alliance with AuNPs for bioagent detection (Shyu et al. 2002). It has also been reported that AuNP coupled with quartz crystal (Jin et al. 2009, Kleo et al. 2011) and SPR (Zhu et al. 2009) are used for bioagent detection.

(c) Quantum Dots (QD)

When two biological entities react there are some significant changes in light emission properties that can be monitored and change recorded provide an excellent detection method for bioagents. The most reasonably applied nanoparticles under these schemes are quantum Dots (QD) carrying some unique properties, like small size, durability and surface functionalization which is flexible and size-reliant emission energy. QD-antibody combination has been successfully utilized for detecting the presence of toxins and viruses like Ricin (Gemmill et al. 2013), *Salmonella typhi* (Yang and Li 2005), *Staphylococcal enterotoxin B* (SEB), Cholera and Shiga toxin (Goldman et al. 2004). QD labeled with antibodies are also reported to be used in the form of fluorescent tags during Flow cytometry (Wang et al. 2005). QD coupled with aptamers has been used in many biosensor systems (Roh and Jo 2011). Biosensor using QD coupled aptamer shows a reduced time of analysis, enhanced sensitivity of detection and amplification of SPR output signal (Anderson et al. 2013).

Sapsford et al. developed a biosensor based on a QD-based BoNT FRET sensor in which the surface of QD was attached with Cy3-labeled peptide substrate. The proteolytic activity of BoNT was monitored by detecting the loss of FRET (fluorescence resonance energy transfer) between QD and Cy3 (Sapsford et al. 2011).

(d) Magnetic Nanoparticles

Magnetic nanoparticles do not display photophysical properties, but in presence of the magnetic field, they can be manipulated and are used in biosensors for purification and concentrating the analytes which reform detection limits and facilitate the procedure of sample preparation (Bromberg et al. 2009). Researchers were able to capture polyanionic DNA from different samples including some environmental samples also. Sequestered genetic material can be identified by PCR (Bromberg et al. 2009).

Biological materials are taken from analytical solutions by nanoparticles which are magnetic by nature. The surface of magnetic nanoparticles coupled with antibodies simplifies the detection process, and there is also no need for a follow-up step of PCR. Magnetic nanoparticles used in biosensor has been demonstrated by Pal and his research group was *Bacillus anthracis* spores can be captured and concentrated by antibody coupled magnetic nanoparticles and further detection through charge transfer mechanism (Pal et al. 2008).

(e) Carbon-Based Nanomaterial

Owing to their biocompatible nature with biomolecules and high electrical conductivity, carbon-based nanomaterials are highly valued and used in biosensors. Any change in emission wavelength is noticed when single-stranded DNA present on the carbon surface hybridizes with an analyte due to its complementary nature, and this indicates the presence of an analyte in solution (Jeng et al. 2006). In 2006, Wang et al. demonstrated that carbon nanotubes can also be used to detect the presence of *B. anthracis* in solution (Wang et al. 2006). Lee with his research team demonstrated

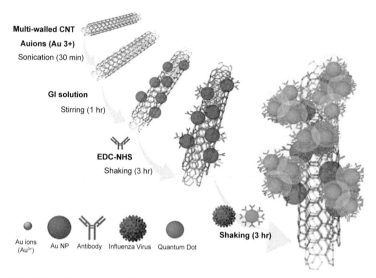

Multi-walled CNT
Auions (Au 3+)
Sonication (30 min)

GI solution
Stirring (1 hr)

EDC-NHS
Shaking (3 hr)

Au ions (Au³⁺)   Au NP   Antibody   Influenza Virus   Quantum Dot

Shaking (3 hr)

Figure 7.3: Schematic representation of synthesis and detection strategy based on CNT/AuNP/QD influenza probes (Reproduced with permission from Lee et al. 2015).

the use of carbon nanotubes as a substrate in a fluorescence-based assay for the development of a composite comprising of carbon nanotube/AuNP, QD for detection of Influenza virus (Lee et al. 2015). In the demonstrated scheme carbon nanotubes were coupled with AuNPs and functionalized with antibodies specifically Influenza antibodies. QD which were antibody functionalized were added to the assembly of CNT/AuNP, which binds with viral analyte captured. The signal given as feedback can be analyzed to detect virus (Figure 7.3).

Single-stranded DNA coupled with a fluorophore on the surface of the C_{60} cluster results in quenched photoluminescence from the fluorophore. On hybridization with analyte DNA, there is the release of DNA/fluorophore complex from C_{60} initiating photoluminescence (Li et al. 2011).

Das et al. designed a composite comprising of carbon nanotube and nanoscale size Zirconiaon an electrode of indium-tin-oxide and demonstrated electrochemical impedance change due to modification of the electrode surface (Das et al. 2011).

(f) Some Newly Emerging Nanomaterials

Some newly developed nanomaterial in recent years has shown promising results when used for the detection of bioagents. One recent example is the use of viruses as nano-scaffold hosting antibodies for the detection and recognition of bioagents and fluorescent dyes for labeling purposes.

The scheme has been effectively used in the detection of SEB and Ricin toxin (Sapsford et al. 2006, Goldman et al. 2009). Recently much research has been conducted on nanomotor which is electrochemically propelled, and they have been demonstrated as an efficient agent in capturing and transporting target species (Orozco et al. 2015).

5. Conclusion

The use of conventional methods for the detection of bioagents has shown excellent sensitivity yet these techniques are time-consuming, very expensive, require well set up laboratory conditions and ample sample preparation for testing purposes. In many conditions, these techniques are insufficient especially in cases where continuous monitoring is required especially during biowarfare or virulent outbreak of diseases in which analysis result is required within a limited time. In such a situation, nonmaterial like QD, carbon nanotubes, AUNPs and magnetic nanomaterial-based biosensors truly excel utilizing nanoscale and dispersibility of nanomaterials to improve the detection and analysis time and for this purpose, many nanomaterial-based biosensors have been developed recently.

References

Anderson, G.P., R.H. Glaven, W.R. Algar, K. Susumu, M.H. Stewart, I.L. Medintz, E.R. Goldman. 2013. Single domain antibody-quantum dot conjugates for ricin detection by both fluoroimmunoassay and surface plasmon resonance. *Anal. Chim. Acta* 786, 132–138.

Baca, O.G., D. Paretsky. 1983. Q-Fever and Coxiella Burnetii—A model for host-parasite interactions. *Microbiol. Rev.* 47, 127–149.

Bearden, S.W., J.D. Fetherston, R.D. Perry. 1997. Genetic organization of the yersiniabactin biosynthetic region and construction of avirulent mutants in Yersinia pestis. *Infect. Immun.* 65, 1659–1668.

Bromberg, L., S. Raduyk, T.A. Hatton. 2009. Functional magnetic nanoparticles for biodefense and biological threat monitoring and surveillance. *Anal. Chem.* 81(14), 5637–5645.

Cao, Y.C., R. Jin, C.A. Mirkin. 2002. Nanoparticles with Raman spectroscopic fingerprints for DNA and RNA Detection. *Science* 297, 1536.

Centers for Disease Control and Prevention. 2018. Bioterrorism Agents/Diseaseswww. emergency.cdc.gov/agent/agentlist-category.asp (accessed August 5, 2020).

Centers for Disease Control and Prevention. 2018. A History of Anthrax, https://www.cdc. gov/anthrax/resources/history/index.html (accessed Apr 10, 2018).

Cheng, X., G. Chen, W. Rodriguez. 2009. Micro- and nano-technology for viral detection. *Anal. Bioanal. Chem.* 393, 487–501.

Chu, M.C. 2001. Basic Laboratory Protocol for the Presumptive Identification of Yersinia pestis; Centers for Disease Control and Prevention: Atlanta, GA.

Condit, R.C. 2001. Principles of virology. *In*: Knipe, D.M., P.M. Howley (eds.). Fields Virology, 4th ed.; Lippincott Williams & Wilkins: Philadelphia.

Croddy, E., S. Krcalova. 2001. Tularemia, biological warfare, and the battle for stalingrad (1942–1943). *Mil. Med.* 166, 837–838.

Das, M., C. Dhand, G. Sumana, A.K. Srivastava, N. Vijayan, R. Nagarajan, B.D. Malhotra. 2011. Zirconia grafted carbon nanotubes based biosensor for *M. Tuberculosis* detection. *Appl. Phys. Lett.* 99, 143702.

de Groot, R.J., S.C. Baker, R.S. Baric, C.S. Brown, C. Drosten, L. Enjuanes, R.A.M. Fouchier, M. Galiano, A.E. Gorbalenya, Z.A. Memish, S. Perlman, L.L.M. Poon, E.J. Snijder, G.M. Stephens, P.C.Y. Woo, A.M. Zaki, M. Zambon, J. Ziebuhr. 2013. Middle East Respiratory Syndrome Coronavirus (MERS-CoV): Announcement of the coronavirus study group. *J. Virol.* 87, 7790–7792.

de Rooij, M.M.T., F. Borlee, L.A.M. Smit, A. de Bruin, I. Janse, D.J.J. Heederik, I.M. Wouters. 2016. Detection of Coxiella burnetii in ambient air after a large Q Fever outbreak. *PLoS One* 11, e0151281.

Dembek, Z.F., J.A. Pavlin, M.G. Kortepeter. 2007. Epidemiology of Biothreat fare and Bioterrorism in Medical Aspects of Biological Warfare, Office of the Surgeon General, US Army Medical Department Center and School: Washington, DC.

Dembek, Z.F. 2011. Medical Management of Biological Casualties Handbook, USAM-RIID.

Dostalova, S., M. Vaculovicova, R. Kizek. 2015. SARS Coronavirus: Minireview. *J. Metallomics Nanotechnol.* 1, 37–42.

Duggan, A.T., M.F. Perdomo, D. Piombino-Mascali, S. Marciniak, D. Poinar, M.V. Emery, J.P. Buchmann, S. Duchêne, R. Jankauskas, M. Humphreys, G.B. Golding, J. Southon, A. Devault, Jean-M. Rouillard, J.W. Sahl, O. Dutour, K. Hedman, A. Sajantila, G.L. Smit, E.C. Holmes, H.N. Poinar. 2016. 17(th) Century variola virus reveals the recent history of smallpox. *Curr. Biol.* 26: 3407–3412.

Feldmann, H. 2014. Ebola a growing threat? *N. Engl. J. Med.* 371, 1375–1378.

Field, L.D., J.B. Delehanty, Y.C. Chen, I.L. Medintz. 2015. Peptides for specifically targeting nanoparticles to cellular organelles: quo vadis? *Acc. Chem. Res.* 48, 1380.

Gemmill, K.B., J.R. Deschamps, J.B. Delehanty, K. Susumu, M.H. Stewart, R.H. Glaven, G.P. Anderson, E.R. Goldman, A.L. Huston, I.L. Medintz. 2013. Optimizing protein coordination to quantum dots with designer peptidyl linkers. *Bioconjugate Chem.* 24(2), 269–281.

Georgiev, V. 2009. Defense Against Biological Weapons (Biodefense), National Institute of Allergy and Infectious Diseases, NIH, Humana Press.

Gill, P., G. Mostafa, G. Amir, P. Gill, M. Ghalami, A. Ghaemi, N. Mosavari, H. Abdul-Tehrani, M. Sadeghizadeh. 2008. Nanodiagnostic method for colorimetric detection of *Mycobacterium tuberculosis* 16S rRNA. *Nanobiotechnology* 4, 28.

Goldman, E.R., A.R. Clapp, G.P. Anderson, H.T. Uyeda, J.M. Mauro, I.L. Medintz, H. Mattoussi. 2004. Multiplexed Toxin analysis using four colors of quantum dot fluororeagents. *Anal. Chem.* 76(3), 684–688.

Goldman, E.R., J.L. Liu, R.D. Bernstein, M.D. Swain, S.Q. Mitchell, G.P. Anderson. 2009. Ricin detection using phage displayed single domain antibodies. *Sensors* 9, 542–555.

Guarise, C., P. Lucia, D.F. Vinsenzo and S. Paolo. 2006. Gold nanoparticles-based protease assay. *Proc. Natl. Acad. Sci. U. S. A.* 103, 3978.

Harkup, K. 2015. A is For Arsenic: The Poisons of Agatha Christie; Bloomsbury Sigma: London, UK.

Hestvik, G., E. Warns-Petit, L.A. Smith, N.J. Fox, H. Uhlhorn, M. Artios, D. Hannant, M.R. Hutchings, R. Mattsson, L. Yon, D. Gavier-Widen. 2015. The status of Tularemia in Europe in a one-health context: A review. *Epidemiol. Infect.* 143, 2137–2160.

Hilgenfeld, R., M. Peiris. 2013. From SARS to MERS: 10 years of research on highly pathogenic human coronaviruses. *Antiviral. Res.* 100, 286–295.

Hussain, M.M., T.M. Samir, H.M.E. Azzazy. 2013. Unmodified gold nanoparticles for direct and rapid detection of *Mycobacterium tuberculosis* complex. *Clin. Biochem.* 46, 633–637.

Islam, M.T., M.A. Uddin. 2017. Biosensors, the emerging tools in the identification and detection of cancer markers. *J. Gynecol. Women's Health* 5(4), 555–667.

Jarocka, U., R. Sawicka, A. Góra-Sochacka, A. Sirko, W. Zagórski-Ostoja, J. Radecki, H. Radecka. 2014. An immunosensor based on antibody binding fragments attached to gold nanoparticles for the detection of peptides derived from avian influenza hemagglutinin H5. *Sensors* 14, 15714–15728.

Jeng, E.S., A.E. Moll, A.C. Roy, J.B. Gastala, M.S. Strano. 2006. Detection of DNA hybridization using the near-infrared band-gap fluorescence of single-walled carbon nanotubes. *Nano Lett.* 6(3), 371–375.

Jin, X., X. Jin, L. Chen, J. Jiang, G. Shen, R. Yu. 2009. Piezoelectric immunosensor with gold nanoparticles enhanced competitive immunoreaction technique for quantification of aflatoxin B1. *Biosens. Bioelectron.* 24, 2580–2585.

Jin, Z., N. Hildebrandt. 2012. Quantum dots for *in vitro* diagnostics and cellular imaging. *Trends in Biotechnology* 30(7), 394–403.

Kaiser, J. 2005. Virology-resurrected influenza virus yields secrets of deadly 1918 pandemic. *Science* 310: 28–29.

Khan, I., L.H. Wieler, F. Melzer, M.C. Elschner, G. Muhammad, S. Ali, L.D. Sprague, H. Neubauer, M. Saqib. 2013. Glanders in animals: A review on epidemiology, clinical presentation, diagnosis and countermeasures. *Transboundary Emerging Dis.* 60, 204–221.

King, A.M., M.J. Adams, E.J. Lefkowitz. 2012. Virus Taxonomy: Classification and Nomenclature of Viruses: Ninth Report of the International Committee on Taxonomy of Viruses; Elsevier: New York.

Kleo, K., A. Kapp, L. Ascher, F. Lisdat. 2011. Detection of vaccinia virus DNA by quartz crystal microbalance. *Anal. Biochem.* 418, 260–266.

Knipe, D.M., P.M. Howley. 2001. Fields Virology: 4th Ed. Lippincott Williams & Wilkins: Philadelphia.

LeClaire, R., M.L. Pitt. 2005. Biological weapons defense: Effect levels. pp. 41–61. *In*: Lindler, L., F. Lebeda, G. Korch (eds.). Biological Weapons Defense; Humana Press: Totowa, NY.

Lee, J., S.R. Ahmed, S. Oh, J. Kim, T. Suzuki, K. Parmar, S.S. Park, J. Lee, E.Y. Park. 2015. A plasmon-assisted fluoro-immunoassay using gold nanoparticle-decorated carbon nanotubes for monitoring the influenza virus. *Biosens. Bioelectron.* 64, 311.

Leroy, E.M., B. Kumulungui, X. Pourrut, P. Rouquet, A. Hassanin, P. Yaba, A. Délicat, J.T. Paweska, J.P. Gonzalez, R. Swanepoel. 2005. Fruit bats as reservoirs of Ebola virus. *Nature* 438, 575–576.

Li, H., Y. Zhang, Y. Luo, X. Sun. 2011. Nano-C_{60}: A novel, effective, fluorescent sensing platform for biomolecular detection. *Small* 7, 1562.

Lin, J.Y., K.Y. Tserng, C.C. Chen, L.T. Lin, T.C. Tung. 1970. Abrin and Ricin: New anti-tumor substances. *Nature* 227, 292–293.

Medintz, I. 2006. Universal tools for biomolecular attachment to surfaces. *Nat. Mater.* 5, 842.

Moczydlowski, E.G. 2013. The molecular mystique of tetrodotoxin. *Toxicon.* 63, 165–183.

Morvan, J.M., V. Deubel, P. Gounon, E. Nakouné, P. Barrière, S. Murri, O. Perpète, B. Selekon, D. Coudrier, A. Gautier-Hion, M. Colyn, V. Volehkov. 1999. Identification of Ebola virus sequences present as RNA or DNA in organs of terrestrial small mammals of the central African Republic. *Microbes Infect.* 1, 1193–1201.

NATO. 1996. Handbook on the Medical Aspects of NBC Defensive Operations, AMedP-6(B); Departments of the Army, the Navy, and the Air Force.

Orozco, J., G. Pan, S. Sattayasamitsathit, M. Galarnyk. 2015. Micromotors to capture and destroy anthrax simulant spores. *Analyst* 140, 14–21.

Owino, J.H., O.A. Arotiba, N. Hendricks, E.A. Songa, N. Jahed, T.T. Waryo, R.F. Ngece, P.G.L. Baker, E.I. Iwuoha. 2008. Electrochemical immunosensor based on polythionine/gold nanoparticles for the determination of aflatoxin B_1. *Sensors* 8, 8262–8274.

Pal, S., E.B. Setterington, E.C. Alocilja. 2008. Electrically active magnetic nanoparticles for concentrating and detecting bacillus anthracis spores in a direct-charge transfer biosensor. *IEEE Sensor J.* 8(6): 647–654.

Patocka, J. 2001. Abrin and Ricin—Two dangerous poisonous proteins. *ASA Newsl.* 85, 20–26.

Reintjes, R., I. Dedushaj, A. Gjini, T.R. Jorgensen, B. Cotter, A. Lieftucht, F.D. Ancona, D.T. Dennis, M.A. Kosoy, G. Mulliqi-Osmani, R. Grunow, A. Kalaveshi, L. Gashi, I. Humolli. 2002. Outbreak investigation in kosovo: Case control and environmental studies. *Emerging Infect. Dis.* 8, 69–73.

Riedel, S. 2004. Biological warfare and bioterrorism: A historical review. *BUMC Proc.* 17, 400–406.

Roh, C., S.K. Jo. 2011. Quantitative and sensitive detection of SARS coronavirus nucleocapsid protein using quantum dots-conjugated RNA aptamer on chip. *J. Chem. Technol. Biotechnol.* 86, 1475–1479.

Saha, K., S.S. Agasti, C. Kim, X. Li, V.M. Rotello. 2012. Gold nanoparticles in chemical and biological sensing. *Chem. Rev.* 112(5), 2739–2779.

Sapsford, K.E., C.M. Soto, A.S. Blum, A. Chatterji, T. Lin, J.E. Johnson, F.S. Ligler, B.R. Ratna. 2006. A cowpea mosaic virus nanoscaffold for multiplexed antibody conjugation: Application as an immunoassay tracer. *Biosens. Bioelectron.* 21, 1668–1673.

Sapsford, K.E., C. Bradburne, J.B. Delehanty, I.L. Medintz. 2008. Sensors for detecting biological agents. *Mater. Today* 11, 38–49.

Sapsford, K.E., J. Granek, J.R. Deschamps, K. Boeneman, J.B. Blanco-Canosa, P.E. Dawson, K. Susumu, M.H. Stewart, I.L. Medintz. 2011. Monitoring Botulinum Neurotoxin A activity with peptide-functionalized quantum dot resonance energy transfer sensors. *ACS Nano.* 5(4), 2687–2699.

Sattarahmady, N., G.H. Tondro, M. Gholchin and H. Heli. 2015. Gold nanoparticles biosensor of *Brucella* spp. genomic DNA: Visual and spectrophotometric detections. *Biochem. Eng. Journal* 97, 1–7.

Schep, L.J., W.A. Temple, G.A. Butt, M.D. Beasley. 2009. Ricin as a weapon of mass terror-separating fact from fiction. *Environ. Int.* 35, 1267–1271.

Shyu, R.H., H.F. Shyu, H.W. Liu, S.S. Tang. 2002. Colloidal gold-based immunochromatographic assay for detection of ricin. *Toxicon.* 40, 255–258.

St. Georgiev, V. 2009. Defense against Biological Weapons (Biodefense). Infectious Disease, Vol. 2: Impact on Global Health; National Institute of Allergy and Infectious Diseases, NIH; Humana Press: Totowa, NY, 221–305.

Stoeva, S.I., J.S. Lee, C.S. Thaxton, C.A. Mirkin. 2006. Multiplexed DNA detection with biobarcoded nanoparticle probes. *Angew. Chem. Int. Ed.* 45, 3303.

Susana Campuzano, S., P.Y. Sedeño, J.M. Pingarrón. 2017. Electrochemical genosensing of circulating biomarkers. *Sensors* 17, 866.

Thiruppathiraja, C., K. Senthilkumar, A. Periyakaruppan, J.S. Devakirubakaran, A. Muthukaruppan. 2011. Specific detection of *Mycobacterium* sp. genomic DNA using dual labeled gold nanoparticle based electrochemical biosensor. *Anal. Biochem.* 417, 73.

Towner, J.S., B.R. Amman, T.K. Sealy, S.A.R. Carroll, J.A. Comer, A. Kemp, R. Swanepoel, C.D. Paddock, S. Balinandi, M.L. Khristova, P.B.H. Formenty, C.G. Albarino, D.M. Miller, Z.D. Reed, J.T. Kayiwa, J.N. Mills, D.L. Cannon, P.W. Greer, E. Byaruhanga, E.C. Farnon, P. Atimnedi, S. Okware, E. Katongole-Mbidde, R. Downing, J.W. Tappero, S.R. Zaki, T.G. Ksiazek, S.T. Nichol, P.E. Rollin. 2009. Isolation of genetically diverse marburg viruses from Egyptian fruit bats. *PLoS Pathog.* 5, e1000536.

Tripathi, K.M., T.Y. Kim, D. Losic, T.T. Tung. 2016. Recent advances in engineered graphene and composites for detection of volatile organic compounds (VOCs) and non-invasive diseases diagnosis. *Carbon* 110, 97–129.

Upadhyayula, V.K.K. 2012. Functionalized gold nanoparticle supported sensory mechanisms applied in detection of chemical and biological threat agents: a review. *Anal. Chim. Acta* 715, 1.

USAMRIID's. 2011. Medical Management of Biological Casulaties Handbook: 7th ed.; Dembek, Z.F. (ed.). U.S. Army Medical Research Institute of Infectious Diseases: Fort Detrick, MD.

Vaseashta, A., E. Braman, P. Susmann. 2012. Technological Innovations in Sensing and Detection of Chemical, Biological, Radiological, Nuclear Threats and Ecological Terrorism, Springer.

Walper, S.A., G.L. Aragonés, K.E. Sapsford, C.W. Brown III, C.E. Rowland, J.C. Breger, I.L. Medintz. 2018. Medintz detecting biothreat agents: from current diagnostics to developing sensor technologies. *ACS Sensors* 3(10), 1894–2024

Wang, L., K.D. Cole, A.K. Gaigalas, Y. Z. Zhang. 2005. Fluorescent nanometer microspheres as a reporter for sensitive detection of simulants of biological threats using multiplexed suspension arrays. *Bioconjugate Chem.* 16(1), 194–199.

Wang, H., L. Gu, Y. Lin, F. Lu, M.J. Meziani, P.G. Luo, W. Wang, L. Cao, Y.P. Sun. 2006. Unique aggregation of anthrax (Bacillus anthracis) spores by sugar-coated single-walled carbon nanotubes. *J. of the American Chemical Society* 128(41), 13364–13365.

Wittig, M.B., P. Wohlsein, R.M. Hagen, S. Al Dahouk, H. Tomaso, H.C. Scholz, K. Nikolaou, R. Wernery, U. Wernery, J. Kinne, M. Elschner, H. Neubauer. 2006. Glanders—a comprehensive review. *Dtsch. Tierarztl. Wochenschr.* 113, 323–330.

World Health Organization. Middle East Respiratory Syndrome Coronavirus (MERS-CoV): Summary of Current Sit-uation, Literature Update and Risk Assessment, WHO/MERS/RA/ 15.1, July 7, 2015; http://www.who.int/csr/disease/coronavirus_infections/risk-assessment-7july2015/en/.

Yang, L.J., Y.B. Li. 2005. Quantum dots as fluorescent labels for quantitative detection of *Salmonella* typhimurium in chicken carcass wash water. *J. Food Prot.* 68: 1241–1245.

Zhou, Q., H.G. Hu, L. Hou. 2015. Discover, develop & validate—advance and prospect of tumor biomarkers. *Clin. Lab.* 61, 1589–1599.

Zhu, S., C.L. Du, Y. Fu. 2009. Localized surface plasmon resonance-based hybrid Au-Ag nanoparticles for detection of Staphylococcus aureus enterotoxin B. *Opt. Mater.* 31, 1608–1616.

Chapter **8**

Nanomaterials-Based Biosensor Application in Environmental Protection

1. Introduction

As a consequence of rapid globalization and industrialization, a broad range of toxic substances is discharged into the environment periodically which causes deterioration of the quality of our natural environment. Some of the environmental pollutants are heavy metals, like Pb^{2+}, Hg^{2+}, Cd^{2+}, As^{3+}, and gases, like SO_2, No_x, etc. Pesticides are released in natural sources, like water bodies, soil and air, due to relevant agricultural activities, industrial and sewage wastewater released directly in water bodies and burning of fossil fuels. Contamination, caused by pesticides and heavy metals, has become a serious menace worldwide since they bring ecological turmoil and accumulate in food chains. Long-term exposure and consumption of water and food adulterated with toxic heavy metals and pesticides can cause certain chronic diseases in humans (Onyido et al. 2004).

The current time demands the development of highly sensitive, specific and economical detection methods and monitoring devices to keep a trail of these toxic substances present in the air, water and soil. The detection technique currently used for monitoring pollutants encompass surface plasmon resonance (Lin et al. 2006, Shankaran et al. 2007), HPCL (Marchetti 1992, Buchheit and Witzenbacher 1996), GC-MS (Gas chromatography-mass spectrometry) (Thurman et al. 1992, Santos and Galceran 2003), ICP-MS (Shum et al. 1992) and many more. All these techniques are time-consuming due to the large size of equipment which requires highly skilled manpower to handle and prepare samples in large quantities. To overcome these constraints of the conventional method, a meticulous biosensing method based on nanomaterial has been developed whose detection limit range from nanomolar to sub-picomolar level (Aragay et al. 2012, Sadik et al. 2009, Banica 2012, Grieshaber et al. 2008, Liu and Lin 2007) and is much more explicit and sensitive.

The application of nanomaterials for the development of analytical biosensors has drawn growing attention in the field of environmental monitoring in recent years owing to their exclusive properties some of which are physiochemical, high adsorption and reaction abilities and high surface-to-volume ratio. Biosensors based on nanomaterial comprise three important components (a) nanomaterial (b) recognition element (c) transducer (Figure 8.1).

(a) Nanomaterials like Quantum Dots, Nanoparticles and Carbon Nanotubes.

(b) Recognition element on which specificity of biosensor depends. A diverse range of recognition elements has been employed in biosensor which includes antibodies (Kim et al. 2011, Liu et al. 2012, Zhao et al. 2012, Trilling et al. 2013, Jiajie et al. 2014), enzymes (Evtugyn et al. 1998), aptamers (Li et al. 2009, Zhang et al. 2011, Li et al. 2011, Long et al. 2013, Ma et al. 2013, Liu et al. 2014), proteins (Bies et al. 2004).

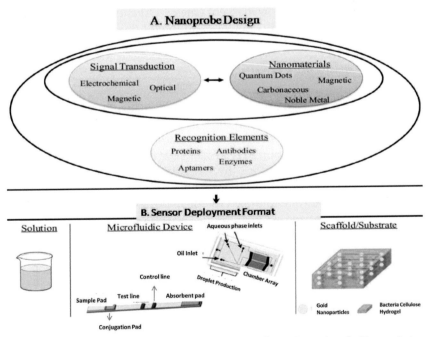

Figure 8.1: A schematic representation of bionanosensor design (reproduced with permission from Willner and Vikesland 2018).

(c) The signal transducer that transduces the signal adopts three methods which are optical, electrochemical, and magnetic.

2. Some Environmental Pollutants Detected by Biosensors

Some most common and highly toxic environmental pollutants, which can be detected by the aid of biosensors are described.

(a) Pesticides

In the present time, there is a great interest among researchers worldwide to develop biosensors for detection of pesticides as they are potent water pollutant is toxic and can bioaccumulate. The pesticides used at the commercial level in the agricultural field comprise over 800 active ingredients present in 100 different substance classes (Liu et al. 2013), which have been presented in Table 8.1. Among the 100 classes of pesticides, brief descriptions of detection techniques for some dominant ones Organophosphorus (QP), carbamates, neonicotinoids and triazine are deliberated.

(i) **Organophosphates**—The mode of action used by pesticides is by targeting certain specific enzymes. Organophosphates target acetylcholinesterase (AChE) enzymes and inhibit their production. AChE is needed during hydrolysis of acetylcholine which is a neurotransmitter (Fukuto 1990, Stenersen 2004).

Yu et al. in 2015 designed an organophosphorus pesticide biosensor utilizing a large surface area of carbon nanotubes (Yu et al. 2015). In the scheme carbon nanotubes (CNT) functionalized with an amino group ($CNT-NH_2$) were dried on the surface of the carbon electrode and afterward incubated with AChE. The estimation limit of this biosensor was around 0.08 nM. It was reported by researchers that nanocomposite can be successfully utilized to improve the stability of AChE biosensors (Cui et al. 2018). In this scheme, reduced graphene oxide (rGO) was stratified on a glassy carbon electrode followed by the deposition of Titanium dioxide (TiO_2) sol-gel film mixed with chitosan. The second layer of chitosan was electro-deposited to improve the stability of the matrix. The estimation limit achieved by this biosensor was reported to be around 29 nM, and detection time required of 25 minutes approximately.

The application of aptamer SERS sensor was explored for the analysis of pesticides in food samples (Pang et al. 2014). In the assay, a unique aptamer developed by Zhang and group (Zhang et al. 2014) was employed, which can detect four different organophosphorus pesticides pharate, profenofos, isocarbophos and omethoate. These aptamers were

Table 8.1: Some common pesticide classes and their effect (Source: Willner and Vikesland 2018).

Class of Pesticides	Example	Types	Effect
Carbamates	Carbaryl, methomyl, propoxur, aldicarb	Fungicide, insecticide,	They are toxic to birds and fishes, non-persistent, cholinesterase-inhibiting
Neonicotinoids	Acetamiprid, clothianidin, imidacloprid, nitenpyram, nithiazine, thiacloprid, thiamethoxam	Insecticide	They are water-soluble and persistence
Organochlorines	Aldrin, chlordane, dieldrin, endrin, heptachlor; lindane, methoxychlor; toxaphene, hexachlorobenzene (HCB), pentachlorophenol (PCP), DDT	Insecticide, acaricide, fungicide	They are persistent and bioaccumulative, affects the endocrine, nervous and immune systems
Organophosphates	Schradan, parathion, malathion	Insecticide, acaricide	They are non-persistent, systemic and toxic to humans
Phenoxy	2,4-D and 2,4,5-T	Herbicide	2,4-D: has the potential to cause cancer in laboratory animals 2,4,5-T: is the source of a toxic contaminant dioxin
Pyrethroids	Fenpropathrin, deltamethrin, cypermethrin	Insecticide	They are toxic to aquatic animals
Triazines	Atrazine, cyanazine, and simazine	Herbicides	They are persistent, binds to the plastoquinone binding protein in photosystem II, endocrine disruptor in humans

stratified on SERS for which dendritic silver was used in form of substrate. Blocking agent 6-mercaptohexanol (MH) was also used whose key role is to eliminate any non-specific type of bonding on the surface of the silver. Probes incubated with pesticides were then evacuated from the solution by centrifugation and were dried preceding Raman interrogation. Raman fingerprint examination of molecules resulted in four distinct detection limit for phorate 0.4 µM, for isocarbophos 3.5 µM, for omethoate 24 µM and profenofos 14 µM. Gold nanoparticles functionalized with antibodies were also utilized to develop an assay named 'bare-eye' (Wang et al. 2014), which enabled visual verification of the presence of three pesticides, chlorpyrifos-methyl and isocarbophos (OP) and imidacloprid.

(ii) Neonicotinoids—First introduced in 1980, neonicotinoids are the largest class of insecticides to be used for agricultural purposes (Jeschke et al. 2011) and are neuro-active. As reported by various researchers, they skeptically affect human health (Simon-Delso et al. 2015). Most of the sensors devised for the detection of neonicotinoids are focused on the detection of acetamiprid with aid of recognition element aptamer (Verdian 2018). It was in 2014 when Weerathunge et al. developed a sensor exploiting aptamer which is based on peroxidase-like activity of gold nanoparticles (Weerathunge et al. 2014). TMB (3,3,5,5-tetramethyl benzidine) was used as a reporter molecule that can change its color to purplish-blue on oxidation. When there is the presence of acetamiprid-specific aptamer, TMB oxidation is blocked as a result it remains colorless. The estimation limit obtained for this scheme was 0.1 ppm.

(iii) Triazine—It belongs to the class of nitrogen-heterocycles. Detection of Triazine is finite to atrazine detection as it is the most widely used herbicide and many biosensors have been devised for its detection. A biosensor was devised in which a gold electrode was embedded with gold nanoparticles and functionalized with anti-atrazine antibodies. The antigen-antibody interaction brings about transformation on the surface of the electrode which was determined with help of pulse voltammetry. The detection limit of this biosensor was 74 pM (Liu et al. 2014).

A nanocomposite of gold nanoparticles and bacteria cellulose was developed on which attachment of atrazine was triggered with pH. The detection limit of this scheme was 11 nM (Wei and Vikesland 2015).

(b) Heavy Metals

The term heavy metals are applied to all metals having a higher density than 5 g/cm^3 and are deliberated highly toxic as even their low concentration poses a serious threat to the natural environment and for the health of humans since they are nonbiodegradable.

Some of the heavy metals that affect our natural environment are cadmium, mercury, lead, arsenic and chromium (Gumpu et al. 2015). The major source of release of heavy metals in water and soil is through the discharge of contaminated water from industries, sewage discharge and agriculture fields in which pesticides are extensively used. The toxicity caused by heavy metals leads to numerous changes at the cellular or molecular level in humans.

The conventional technology used for the analysis of heavy metals present in the natural environment includes atomic absorption spectrometry, inductively coupled plasma mass spectrometry and X-ray fluorescence spectrometry, but all these techniques are time-consuming and require large sample preparation. As these metals cause severe toxicity the need for the hour is to develop a biosensor that can analyze the presence of these metals within limited time and high precision.

(i) Mercury (Hg)—The source of incorporation of mercury in our natural environment is by combustion of fossil fuels, discharge from industrial and agricultural sites. They are considered highly poisonous for humans (Cargnelutti et al. 2006). For the detection of mercury, Liu et al. devised a special biosensor and mercury sandwich assay. In the scheme described by him, magnetic silica spheres were enclosed in a gold shell and Raman labeled gold nanoparticles were functionalized with DNA sequences which were analogous to it and contain five incongruous thymine site. In presence of mercury, hybridization occurs which diminished the inter-probe spacing and creates a plasmonic hotspot. Nanoprobes can be recovered easily due to their magnetic core with help of a magnet (Liu et al. 2014).

Another assay for the detection of mercury was reported which is mediated by Thiol. In the assay, various nanoparticles are used such as gold (Huang and Chang 2006, Kim et al. 2009, Xiaorong et al. 2011, Chansuvarn et al. 2015), silver (Alam et al. 2015) and quantum dots (Ke et al. 2012). Aggregation of molecules is utilized for the generation of calorimetric responses (Kim et al. 2009).

A fluorescent signal emitting biosensor in presence of mercury was developed by Huang and Chang. In presence of mercury, there is a displacement of Rhodamine 6G from the surface of the nanoparticles. The sensor has a detection limit of 2.0 ppb and an analysis timing of fewer than 10 minutes (Huang and Chang 2006).

(ii) Lead (Pb)—They are found in nature in form of sulfides, sulfates and carbonates and the main anthropogenic source of this environmental pollutant in nature is fossil fuel combustion, fire at landfills, industrial and agricultural waste disposal. They are highly toxic for humans as they are briskly absorbed in the bloodstream and immediately binds to RBC in Pb^{2+} form and finally get deposited in bones in $Pb_3(PO4)_2$ form and affect our nervous and immunological system.

For the detection of Pb^{2+}, many biosensors have been developed. Tang and his research group developed an electrochemical biosensor combining 8-17 DNAzyme, rolling circle amplification (RCA) and quantum dots (Tang et al. 2013). In the assay catalytic strand of DNAzyme was developed which was immobilized on a magnetic bead and hybridized on a substrate containing ribonucleoside adenosine (rA), sessile in nature to form double-stranded DNA thatholds Pb^{2+} ions. Pb^{2+} ion presence activates DNAzyme activity and a nick was created at rA group present in the substrate. Single DNA strand now exposed was restrained to magnetic beads and then hybridized with RCA template to obtain a long single strand repeating sequence. The analogous sequence of RCA was functionalized to CdS quantum dots leading to the hybridization of multiple QDs. QDrich DNA strand was separated magnetically and dissolved in a solution of nitric acid which led to the release of cadmium cation detected by square wave voltammetry. The estimation limit achieved for this biosensor was 7.8 pM.

(iii) Chromium (Cr)—In form of a compound, chromium exists in various oxidation states from bivalent to hexavalent ion. In aqueous form, chromium can exist in trivalent and hexavalent forms. Cr(VI) form of chromium is toxic and carcinogenic and enters the biological system easily crossing the biomembrane and has a high degree of oxidation. When water and food contaminated with Cr(VI) are consumed, it causes severe damage to our liver, causes lung cancer and skin irritation and ulcer formation (Cobatingan et al. 2001) and tumors (Gibb et al. 2000).

Many immunoassays have been developed for the detection of Cr (Liu et al. 2012, Jiajie et al. 2014) based on work done by Liu and his research team. In the assay, developed CR was mixed with bifunctional chelating agent Isothiocyanobenzyl-EDTA. Due to its small size, it cannot bring out any immune response. Later on, they were adjoined with bovine serum albumin (BSA) acting as a carrier protein and then introduced to mice. The immunoassay comprises of three parts (1) conjugation pad embedded with anti-Cr-EDTA antibodies, (2) analyte of interest Cr-EDTA (test line) and (3) goat-anti-mouse antibodies (control line). In the scheme, a liquid solution was introduced which travels up to the conjugation pad where probes are brought into the solution.

In a negative sample, the free antibodies probe binds to the test line and for a positive sample there is no binding of the probe as all antibody sites are already invaded and hence no signal will be produced. The estimation limit achieved for this assay was around 50 ng/mL.

3. Conclusion

In the current scenario, biosensors have found a wide range of applications in various fields, including monitoring pollutants in the environment. The efficiency of a biosensor for better and specific detection of environmental pollutants like pesticides and heavy metals has been greatly enhanced by the use of several nanomaterials. The aim of the recent development in designing a biosensor with aid of nanomaterials is to overcome the limitation of currently used techniques and has shown a great potential for tracing and detecting pollutants.

References

Alam, A., A. Ravindran, P. Chandran, S.S. Khan. 2015. Highly selective colorimetric detection and estimation of Hg2+ at nano-molar concentration by silver nanoparticles in the presence of glutathione. *Spectrochim. Acta Part A Mol. Biomol. Spectrosc.* 137, 503–508.

Aragay, G., F. Pino and A. Merkoci. 2012. Nanomaterials for sensing and destroying pesticides. *Chem. Rev.* 112, 5317–5338.

Bănică, F.-G. 2012. Nanomaterial applications in optical transduction. pp. 454–472. *In*: Chemical Sensors and Biosensors. Chichester: Wiley.

Bies, C., C.M. Lehr, J.F. Woodley. 2004. Lectin-mediated drug targeting: history and applications. *Adv. Drug Deliv. Rev.* 56, 425–35.

Buchheit, A.J., M. Witzenbacher. 1996. Pesticide monitoring of drinking water with the help of solid-phase extraction and high-performance liquid chromatography. *J. Chromatogr. A* 737, 67–74.

Cabatingan, L.K., R.C. Agapay, J.L.L. Rakels, M. Ottens, L.A.M. van der Wielen. 2001. Potential of biosorption for the recovery of chromate in industrial wastewaters. *Ind. Eng. Chem. Res.* 40, 2302–2309.

Cargnelutti, D., L.T. Almeri, R.M. Spanevello, G. de O. Jucoski, V. Battisti, M. Redin, C.E.B. Linares, V.L. Dressler, É.M. de M. Flores, F.T. Nicoloso, V.M. Morsch, M.R.C. Schetinger. 2006. Mercury toxicity induces oxidative stress in growing cucumber seedlings. *Chemosphere* 65, 999–1006.

Chansuvarn, W., T. Tuntulani, A. Imyim. 2015. Colorimetric detection of mercury(II) based on gold nanoparticles, fluorescent gold nanoclusters and other gold-based nanomaterials. *TrAC Trends Anal. Chem.* 65, 83–96.

Corcia, A.D., M. Marchetti. 1992. Method development for monitoring pesticides in environmental waters: liquid-solid extraction followed by liquid chromatography. *Environ. Sci. Technol.* 26, 66–74.

Cui, H.F., W.W. Wu, M.M. Li, X. Song, Y. Lv, T.T. Zhang. 2018. A highly stable acetylcholinesterase biosensor based on chitosan-TiO2-graphene nanocomposites for detection of organophosphate pesticides. *Biosens. Bioelectron.* 99, 223–229.

Evtugyn, G.A., H.C. Budnikov, E.B. Nikolskaya. 1998. Sensitivity and selectivity of electrochemical enzyme sensors for inhibitor determination. *Talanta* 46, 465–484.

Fukuto, T.R. 1990. Mechanism of action of organophosphorus and carbamate insecticides. *Environ. Health Perspect.* 87, 245–254.

Gibb, H.J., P.S. Lees, P.F. Pinsky, B.C. Rooney. 2000. Lung cancer among workers in chromium chemical production. *Am. J. Ind. Med.* 38, 115–126.

Grieshaber, D., R. MacKenzie, J. Vörös, E. Reimhult. 2008. Electrochemical biosensors—sensor principles and architectures. *Sensors* 8, 1400–1458.

Gumpu, M.B., S. Sethuraman, U.M. Krishnan, J.B.B. Rayappanabd. 2015. A review on detection of heavy metal ions in water— An electrochemical approach. *Sens. Actuators B.* 213, 515–533.

Huang, C.C., H.T. Chang. 2006. Selective gold-nanoparticle-based "turn-on" fluorescent sensors for detection of mercury(II) in aqueous solution. *Anal. Chem.* 78, 8332–8338.

Jeschke, P., R. Nauen, M. Schindler, A. Elbert. 2011. Overview of the status and global strategy for neonicotinoids. *J. Agric. Food Chem.* 59, 2897–2908.

Jiajie, L., L. Hongwu, L. Caifeng et al. 2014. Silver nanoparticle enhanced Raman scattering-based lateral flow immunoassays for ultra-sensitive detection of the heavy metalchromium. *Nanotechnology* 25, 495–501.

Ke, J., X. Li, Y. Shi, Q. Zhaoa, X. Jiang. 2012. A facile and highly sensitive probe for Hg(II) based on metal-induced aggregation of ZnSe/ZnS quantum dots. *Nanoscale* 4, 4996–5001.

Kim, Y.A., E.H. Lee, K.O. Kim, Y.T. Lee, B.D. Hammock, H.S. Lee. 2011. Competitive immunochromatographic assay for the detection of the organophosphorus pesticide chlorpyrifos. *Anal. Chim. Acta* 693, 106–113.

Kim, Y.R., R.K. Mahajan, J.S. Kim, H. Kim. 2009. Highly sensitive gold nanoparticle based colorimetric sensing of mercury(II) through simple ligand exchange reaction in aqueous media. *ACS Appl. Mater. Interfaces* 2, 292–5.

Li, M., Q. Wang, X. Shi, L.A. Hornak, N. Wu. 2011. Detection of mercury(II) by quantum dot/DNA/gold nanoparticle ensemble based nanosensor viananometal surface energy transfer. *Anal. Chem.* 83, 7061–7065.

Li, T., B. Li, E. Wang, S. Dong. 2009. G-quadruplex-based DNAzyme for sensitive mercury detection with the naked eye. *Chem. Commun.* https://doi.org/10.1039/b9039 93g.

Lin, T., K. Huang, C. Liu. 2006. Determination of organophosphorous pesticides by a novel biosensor based on localized surface plasmon resonance. *Biosens. Bioelectron.* 22, 513–518.

Liu, G., Y. Lin. 2007. Nanomaterial labels in electrochemical immunosensors and immunoassays. *Talanta* 74, 308–317.

Liu, M., Z. Wang, S. Zong, H. Chen, D. Zhu, L. Wu, G. Hu, Y. Cui. 2014. SERS detection and removal of mercury(II)/silver(I) using oligonucleotide-functionalized core/shell magnetic silica Sphere@Au nanoparticles. *ACS Appl. Mater. Interfaces* 6, 7371–7379.

Liu, S., Z. Zheng, X. Li. 2013. Advances in pesticide biosensors: current status, challenges, and future perspectives. *Anal. Bioanal. Chem.* 405, 63–90.

Liu, X., J.J. Xiang, Y. Tanga, X.L. Zhang, Q.Q. Fu, J.H. Zou, Y.H. Lin. 2012. Colloidal gold nanoparticle probe-based immunochromatographic assay for the rapid detection of chromium ions in water and serum samples. *Anal. Chim. Acta* 745, 99–105.

Liu, X., W.J. Li, L. Li, Y. Yang, L.G. Mao, Z. Peng. 2014. A label-free electrochemical immunosensor based on gold nanoparticles for direct detection of atrazine. *Sens Actuators B Chem.* 191, 408–414.

Long, F., A. Zhu, H. Shi, H. Wang, J. Liu. 2013. Rapid *on-site/in-situ* detection of heavy metal ions in environmental water using a structure-switching DNA optical biosensor. *Sci. Rep.* 3, 2308.

Ma, J., Y. Chen, Z. Hou, W. Jiang, L. Wang. 2013. Selective and sensitive mercuric(II) ion detection based on quantum dots and nicking endonuclease assisted signal amplification. *Biosens. Bioelectron.* 43, 84–87.

Onyido, A., R. Norris, E. Buncel. 2004. Biomolecule–mercury interactions: Modalities of DNA base–mercury binding mechanisms. *Remediation Strategies Chem. Rev.* 104, 5911–5930.

Pang, S., T.P. Labuza, L. He. 2014. Development of a single aptamer-based surface enhanced Raman scattering method for rapid detection of multiple pesticides. *Analyst* 139, 1895–1901.

Sadik, O.A., A.O. Aluoch, A. Zhou. 2009. Status of biomolecular recognition using electrochemical techniques. *Biosens. Bioelectron.* 24, 2749–2765.

Santos, F.J. and M.T. Galceran. 2003. Modern developments in gas chromatography–mass spectrometry-based environmental analysis. *J. Chromatogr. A* 1000, 125–151.

Shankaran, D.R., K.V. Gobi, N. Miura. 2007. Recent advancements in surface plasmon resonance immunosensors for detection of small molecules of biomedical, food and environmental interest. *Sens. Actuat. B* 121, 158–177.

Shum, S.C.K., H.M. Pang, R.S. Houk. 1992. Speciation of mercury and lead compounds by microbore column liquid chromatography-inductively coupled plasma mass spectrometry with direct injection nebulization. *Anal. Chem.* 64, 2444–2450.

Simon-Delso, N., R.V. Amaral, L.P. Belzunces, J.M. Bonmatin, M. Chagnon, C. Downs, L. Furlan, D.W. Gibbons, C. Giorio, V. Girolami, D. Goulson, D.P. Kreutzweiser, C.H. Krupke, M. Liess, E. Long, M. McField, P. Mineau, E.A.D. Mitchell, C.A. Morrissey, D.A. Noome, L. Pisa, J. Settele, J.D. Stark, A. Tapparo, H. Van Dyck, J. Van Praagh, J.P. Van der Sluijs, P.R. Whitehorn, M. Wiemers. 2015. Systemic insecticides (neonicotinoids and fipronil): trends, uses, mode of action and metabolites. *Environ. Sci. Pollut. Res. Int.* 22, 5–34.

Stenersen, J. 2004. Chemical Pesticides Mode of Action and Toxicology. Boca Raton: CRC Press.

Tang, S., P. Tong, H. Li, J. Tang, L. Zhang. 2013. Ultrasensitive electrochemical detection of Pb(2)(+) based on rolling circle amplification and quantum dots tagging. *Biosens. Bioelectron.* 42, 608–611.

Thurman, E.M., D.A. Goolsby, M.T. Meyer, M.S. Mills, M.L. Pomes, D.W. Kolpin. 1992. A reconnaissance study of herbicides and their metabolites in surface water of the midwestern United States using immunoassay and gas chromatography/mass spectrometry. *Environ. Sci. Technol.* 26, 2440–2447.

Trilling, A.K., J. Beekwilder, H. Zuilhof. 2013. Antibody orientation on biosensor surfaces: a minireview. *Analyst* 138, 1619–1627.

Verdian, A. 2018. Apta-nanosensors for detection and quantitative determination of acetamiprid—a pesticide residue in food and environment. *Talanta* 176, 456–464.

Wang, L.M., J. Cai, Y.L. Wang, Q. Fang, S. Wang, Q. Cheng, D. Du, Y. Lin, F. Liu. 2014. A bare-eye-based lateral flow immunoassay based on the use of gold nanoparticles for simultaneous detection of three pesticides. *Microchim. Acta* 181, 1565–1572.

Weerathunge, P., R. Ramanathan, R. Shukla, T.K. Sharma, V. Bansal. 2014. Aptamer controlled reversible inhibition of gold nanozyme activity for pesticide sensing. *Anal. Chem.* 86, 11937–11941.

Wei, H., P.J. Vikesland. 2015. pH-triggered molecular alignment for reproducible SERS detection via an AuNP/nanocellulose platform. *Sci. Rep.* 5, 18131.

Willner, M.R., P.J. Vikesland. 2018. Nanomaterial enabled sensors for environmental contaminants. *J. Nanobiotechnol.* 16, 95.

Xiaorong, Y., L. Huixiang, X. Juan, T. Xuemei, H. Huang, T. Danbi. 2011. A simple and cost-effective sensing strategy of mercury(II) based on analyte inhibited aggregation of gold nanoparticles. *Nanotechnology* 22, 275503.

Yu, G., W. Wu, Q. Zhao, X. Wei, Q. Lu. 2015. Efficient immobilization of acetylcholinesterase onto amino functionalized carbon nanotubes for the fabrication of high sensitive organophosphorus pesticides biosensors. *Biosens. Bioelectron.* 68, 288–294.

Zhang, C., L. Wang, Z. Tu, X. Sun, Q. He, Z. Lei, C. Xu, Y. Liu, X. Zhang, J. Yang, X. Liu, Y. Xu. 2014. Organophosphorus pesticides detection using broad-specific single-stranded DNA based fluorescence polarization aptamer assay. *Biosens. Bioelectron.* 55, 216–219.

Zhang, M., B.C. Yin, W. Tan, B.C. Ye. 2011. A versatile graphene-based fluorescence "on/off" switch for multiplex detection of various targets. *Biosens. Bioelectron.* 26, 3260–3265.

Zhao, W.W., Z.Y. Ma, P.P. Yu, X.Y. Dong, J.J. Xu, H.Y. Chen. 2012. Highly sensitive photoelectrochemical immunoassay with enhanced amplification using horseradish peroxidase induced biocatalytic precipitation on a Cd Squantum dots multilayer electrode. *Anal. Chem.* 84, 917–923.

Index